Cyclopedic Treasury of Arts and Crafts Activities Using Scrap Materials

Berlyn Erdahl

Parker Publishing Company, Inc.
West Nyack, N.Y.

To Shirley
for storing many "treasures"
and for reading these many projects over and over.
And to collectors of "treasured scraps" everywhere.

Library of Congress Cataloging in Publication Data

Erdahl, Berlyn J
Cyclopedic treasury of arts and crafts activities using scrap materials.

Includes index.
1. Handicraft. 2. Creative activities and seat work. 3. Waste products. I. Title.
TT157.E7   745.5   77-7056
ISBN 0-13-196600-6

Printed in the United States of America

Foreword

This volume deals with a fascinating array of ideas for creative expression through the use of materials that commonly find their way into the discard category. But now "Don't throw that away!" has become the slogan of many households. In these days of concern about the pollution of our environment, the recycling of materials is a topic worthy of consideration in itself. Art projects utilizing otherwise useless items can thus serve a dual purpose as ecological concerns are met through a satisfying, artistic experience.

Throughout the *Cyclopedic Treasury of Arts and Crafts Activities Using Scrap Materials* you will find references to the integration of art with other curricular areas. As children make magazine mosaics, they can learn about the art forms of ancient civilizations. Searching for the materials for "Rub Out the Weeds" will provide an opportunity to identify and classify plants. Papier-mâché rhythm instruments made by the children will be used much more enthusiastically than the commercial variety. Oral language activities tie in with making puppets, masks, and TV sets.

Many of the projects included in this book have been tried over the years by many teachers and literally thousands of children. Use them in your classroom or outside of school with children's groups and be ready to experience the satisfaction of seeing a child's face as he exclaims, "Look what *I* made!"

Dr. Lois Stephens
Director of Instructional Services
Oxnard School District
Oxnard, California

Introduction

What can you do with all of those cardboard tubes? Or styrofoam packing? Or magazines? Or plastic lids? Are you a saver? A hoarder? A pack rat? If you are, as teachers are prone to be because of budget limitations or because you are being mindful of ecology, and you suddenly realize that you are being overrun with discards, this book is for you. It is aimed at being a ready reference for the use of scrap materials and "found" objects in the elementary school art program. Making "things" is only one part of it and using scrap materials is only one approach to an art curriculum. All parts of the art program have something special to offer the child in stimulating his visual awareness, which is, after all, our major concern.

Everyday materials used in new ways lead to seeing in new ways. The crayon and cut-and-paste experience is strong, positive, and valid, but a little variety in any diet is welcomed by most and demanded by many. Perhaps the child who responds passively to the usual materials will really "turn on" to the unusual. Maybe the break from tradition is what the bored child needs to recharge his creative spark and what the already inspired child needs to expand. The teacher is, of course, the energy force that ignites this creative spark for inventive use of any materials in any part of the curriculum.

Enthusiastic motivation on the part of an inspired teacher can carry the momentum necessary to stimulate exploration of simple but intriguing materials. As in all art experiences, there are no predetermined answers. There is latitude and freedom within the limits of the problem, allowing for as many solutions as there are problem-solvers. A built-in advantage to scrap materials, which encourages inventive, unusual, and even oddball solutions to the problems, is that they are being changed and given a second chance—a new, recycled use.

The suggested activities in this book are more than a bag-of-tricks. They are not simple formulae, but vehicles through which the child can communicate with himself and with others. Through them, the teacher can lead the child to develop an attitude of respect toward craftsmanship as well as a curiosity and innovative response to the materials at hand, whether it be sophisticated art or non-art. Basic material-centered, how-to ideas are presented with the expectation that the teacher will pursue individual development of the suggestion. A young child comes to school with a spontaneous creative attitude that allows him to assemble many kinds of materials with an innate sense of design. We can hope to extend this attitude to later years through motivation. Insight and the ability for personal observations are natural outgrowths of this getting together with materials.

Self-identity and the feeling of importance in a world of impersonal mechanization and computerization can be fostered through searching and finding solutions to creative problems with scrap materials. The self-concept can be greatly elevated with the feeling of accomplishment and success from making something of nothing or just changing its character. A what-is-it or what-is-it-for or what-does-it-mean is not always necessary.

Introduction

Just an object for its own sake and the fact that it was done by a child who will have a sense of "it's mine, I made it" accomplishment is enough reason for being. The teacher can share in this experience by providing the motivation and approval when it is completed.

As with all education, we want the child to carry his experiences beyond the walls of the classroom into the home and through life after the school years. The joy of creating is a human experience necessary for a rich life and can be nurtured at school and perpetuated through attitude into adulthood and leisure time. The child's involvement in the search for and collection of scrap materials will develop a sensitivity to art elements. Line, form, color, and texture will become something real in his everyday environment, rather than abstract ideas pointed out only in a reproduction of Picasso's *Three Musicians.*

The self-contained classroom teacher is in a unique position to help develop children's interest in meaningful activities and concepts, including the fact that art is a part of a total life style. As the student discovers found objects and new uses for them, he will also be learning to see with his fingers and feel with his eyes. His visual literacy will expand as the doors open to discovering and rediscovering what is close at hand. One man's trash is another man's treasure will become a concept in fact for him.

—*Berlyn Erdahl*

Contents

Contents

Contents

Contents

Contents

Contents

Candles
&
Corks

Wood Scraps & Sawdust

Wood Scraps with Crayon and Shellac

A collection of odd-shaped pieces of wood is the starting point for an imaginary animal or a useful keyholder. Browsing through the varied shapes and sizes may suggest a long-necked or large-eared creature that needs only some line and color to make it come alive. Or the shape may simply be a form that is pleasant to look at and handle. With some imagination, crayons, and the addition of cup hooks or nails, it will be a handy collector for the keys that sometimes get lost.

A gift for dad? A gift for mom to hold her measuring spoons? Let's go!

MATERIALS

wood scraps
wax or oil crayons
shellac
brushes
sandpaper
alcohol (to clean brushes)

PROCEDURE

Prepare your selected piece of wood by sanding the edges with fine sandpaper. Turn the wood several times to find the angle most suggestive of an animal or other shape.

The crayon work begins with a light sketch (to determine color areas) so that changes can be made easily as your thinking progresses. The final colors should be applied very heavily to create a strong color surface. A layer of wax or oil (depending upon the kind of crayons used) is desired, rather than simply a coating of pigment.

Finishing is done with a coat or two of shellac. Be sure to brush the shellac out into an even coating. If a second coat is desired, allow the first to dry completely before proceeding. Brushes should be cleaned in alcohol and then soap and water.

If the children are making animals for a zoo scene, both sides of the wood should be crayoned. A scrap of wood might be glued to the bottom as a stand. In the case of a key or spoon holder, screw in cup hooks or drive nails across the bottom edge. Add a hanger to the back. This hanger could be made from a pull-top ring. Simply glue, tape, or staple it on. A welcome gift!

Wrap It and Print It

Abstract line designs to be printed on a folder, wrapping paper, or fabric may first be conceived on a scrap of wood. The wood scrap box serves us once again! This time the smaller pieces will be put to work.

MATERIALS

scrap wood
twine or carpet warp
white glue
brushes
scissors
tempera or textile paint
newspapers

PROCEDURE

Wrap the piece of wood with twine, keeping it taut and overlapping the line to develop a design. Think of the twine as a line that might be drawn with a pencil. When a satisfying design has been completed, glue the ends of the twine on the *back* of the block using a small amount of white glue. Allow the glue to dry and the block is ready for printing.

For printing on paper, thin the tempera paint with water. For fabric, use a textile paint, preferably a water-soluble one since cleanup is easier.

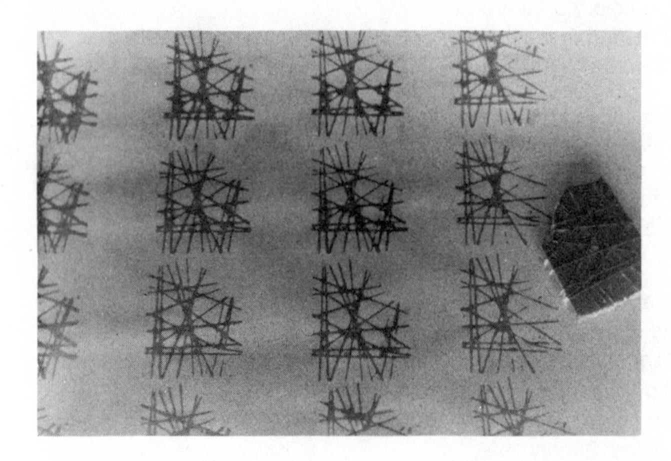

Wrap It and Print It

Apply the paint to the twine with a brush until it is well saturated with color. Don't be concerned about getting a little on the wood block, since the twine is raised. As long as it is well saturated with paint, several impressions can be made with the printing block before re-inking is necessary. A padding of several layers of newspaper under the printing surface will create a cushion that helps make a sharper print.

Printing patterns may be done in regular or random repeats as shown in the illustration.

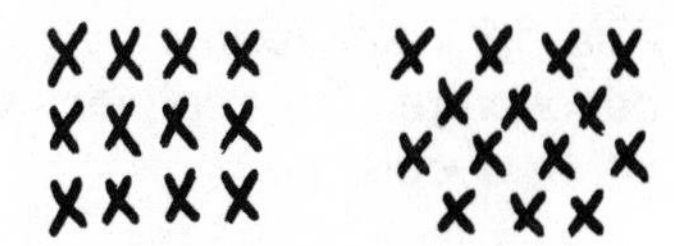

RANDOM REPEAT

Découpage Plaque

A favorite picture in a magazine can be preserved in a decorative wall piece called découpage. This treatment will make a simple picture into an "antiqued" treasure for a gift or decoration.

MATERIALS

hammer
scraps of wood
sandpaper
sponges
white glue
magazine pictures (slick, colored)
black or brown tempera or acrylic paint
shellac

PROCEDURE

Select a magazine picture. Pick a piece of wood larger than the picture and sand the edges. "Distress" the wood surface and edges by pounding them with a hammer. If the pounding produces any splinters, they should be sanded off to give a finished look to the piece.

Tear the picture so there is a feathered rather than a sharp, hard edge. Place it on the board near the center to make sure that it will fit. Brush out the white glue in a smooth coating on the wood, then lay the picture on the wet glue. Smooth it out with the brush dipped in glue. Avoid scrubbing it or the color on the picture will smear. Allow this to dry completely.

Antiqued Surface. Dip a wet sponge into the dark tempera paint and wipe it on the wood and across the picture from top to bottom. With a clean sponge that is slightly damp wipe across the surface to clean away a small amount of paint. Again, work from top to bottom with light pressure to avoid damage to the picture surface. Keep working until the desired light and dark appearance of age is achieved and until the picture is clear enough. Rinse the sponge when it gets too full of paint to do any cleaning. Allow this to dry completely.

Apply a coat or two of shellac (or acrylic medium) for a finished surface.

Finally, add a hanger to the top of the plaque as shown in the example. If decorative hangers are unavailable, use a simple pull-top ring on the back of the wood.

Panel Assemblage

An assemblage is, as the name suggests, a composition assembled from many pieces. It is composed of "what's at hand," whether they are normally art materials or not. We can make use of small and large scraps of wood to create a panel of interest in shapes, textures, and lights and darks. A well-known artist, Louise Nevelson, used wooden forms, often scraps, to make large walls of exciting, eye-appealing relief patterns.

MATERIALS

wood panel (or heavy cardboard)
scraps of wood
sandpaper
white glue

PROCEDURE

With a large selection of sizes and shapes of wood scraps, begin to arrange and rearrange designs on a large piece of wood or very heavy cardboard. The total size of the finished piece is almost unlimited, but the larger the piece the stronger the base will have to be. Any piece that has very rough edges should be sanded to avoid splinters and an unfinished look. Don't be too easily satisfied with the arrangement of the pieces. Keep changing them around until you feel that the best design has been achieved. Overlapping pieces will create a feeling of unity, but will keep the design in relief (only slightly raised from the background).

Then, using white glue, attach all of the pieces and allow them to dry. A hanger on the back completes the assemblage. Each child could make his or her own assemblage and then all of them could be joined together for a giant mural.

The wood may be stained or painted a single color (e.g., white, black, or brown), but avoid multicolor painting. Let the light and dark of the shadows from the high-low surface create interest.

Alternative: Use the wooden high-low shapes as a background for pictures collected from magazines. In the pictured example heads have been used, but any subjects would work if they are similar or related in some way. For example, flowers or animals might be fun. Glue them on with white glue and brush over them with more glue (dries clear) or shellac.

Panel Assemblage

Sawdust Sculpture

Wall-decoration sculpture in high relief can be modeled of sawdust. The sawdust is combined with plaster to make a material similar to pressed wood. A sawdust picture with a third dimension is exciting and easily achieved with this material.

MATERIALS

sawdust (fine)
white glue
plaster
cardboard
tempera paint
brushes

PROCEDURE

Mix several cups of fine sawdust with a little dry plaster (approximately ½ cup of plaster to 2 cups of sawdust). It is important to mix the dry ingredients before adding any liquid. Now add diluted white glue, about ½ water. Mix with wet hands by squeezing and turning until it becomes a firm, manageable mass. Add more glue mixture if it doesn't stick together. Place this on a piece of cardboard which is a little larger than what the sculpture will be. Faces are a good subject because they are relatively tight masses (they don't have protruding parts). This is a relief sculpture that rises from the background, thus gaining support from the cardboard. Keep the fingers damp to prevent sticking during the modeling process. Press and pinch the shapes to ensure a strong finished piece. Other tools, such as popsicle sticks, spoons, pencils, etc., may be used. When finished with the form, set it aside to dry at least overnight. With care, it may be set into an oven at very low temperature to "force dry," but some warpage may occur.

The surface may be sanded with fine sandpaper as you would sand a piece of wood, but it must be completely dry first.

Paint the finished piece a solid color, preferably a very dark one. After it is dry, use a dry-brush technique to achieve an "antique" look and to bring out the highlights by creating light and dark patterns. To get a dry brush, dip your brush in paint (a light color) and wipe most of the paint off on newspapers. Then, quickly brush over the surface hitting the highest points in several directions.

An alternative is to paint the piece with several colors, creating a multicolored decorative effect resembling a mask-like head. Other materials might be glued on, such as yarn, fur, hair, or feathers.

Attach a hanger to the back for a wall decoration. A pull-top ring makes an ideal hanger for this piece, and it costs nothing!

Nonobjective Wood Sculpture

You understand three-dimensional space more clearly when you actually create the space in and around a form, using scraps of wood much as small children use building blocks. However, with wood scraps you aren't limited to the cube, but have a wide range of shapes limited only by your scrap supply. Building up and out with very few limits on dimension becomes an exciting problem in form and space.

There is no need to make this design look like a person, an animal, or an object. The "what is it" of this sculpture is unimportant. We are not concerned with representing a particular thing, but rather with creating a whole new idea—a nonobjective sculpture.

MATERIALS

wood scraps
white glue
sandpaper

PROCEDURE

Beginning the sculpture with a strong base is quite important to the stability of the completed piece. Use sufficient glue to hold the pieces together as you build. Too much glue, however, retards drying and really doesn't add to the strength. Hold each piece in place until it seems secure. Balancing one piece against another, even temporarily, will help. A *small* piece of masking tape will be helpful in holding some pieces together until the glue dries.

Consideration should be given to strength as well as design when supporting shapes with other shapes. Since this is a sculpture, it is advisable to keep working around the piece as it grows, so all sides will be developed.

The natural colors of the wood can be retained (as in the pictured example), or it could be painted or stained.

Toothpick Constructions

Construct a three-dimensional line drawing. A drawing in three dimensions? It could be considered that, since each toothpick is like a straight line in a geometric drawing that can extend in many directions. The final form need not be predetermined, but rather allow it to grow and develop as you work. A constant concern for space and balance must be maintained, however, in order to have a strong construction and yet a sculptural form with eye appeal. This is an enjoyable experience for children of all ages, with differing degrees of sophistication. Regardless of whether the child creates something simple or complex, the construction helps him develop a clearer understanding of space and balance.

MATERIALS

toothpicks
quick-drying household cement (e.g., Duco)
newspapers

PROCEDURE

Begin by making a base on which to build, such as a square or triangle of toothpicks. Use small amounts of glue at the tips of the toothpicks so they will dry very quickly. Each one can be dipped in the glue as they are added to the sculpture, or one toothpick may be used as an applicator. The technique used is a matter of personal preference, but small amounts of glue make the process easier. Blow gently to "force dry" any that seem to be drying too slowly. Keep the table protected from glue with newspapers. Now build up, out, and away!

Working on a small piece of cardboard or a plastic lid from a coffee can will allow you to turn the form as you work. Keep rotating it so that it can be seen from all sides. Consider balance and the direction of the lines—as in drawing with straight lines in many directions at once. The use of large and small shapes within the sculpture will make the finished piece more interesting.

Children can creatively design toothpick animals, boats, buildings, or whatever. Or they can simply experiment in three-dimensional form as shown in the example.

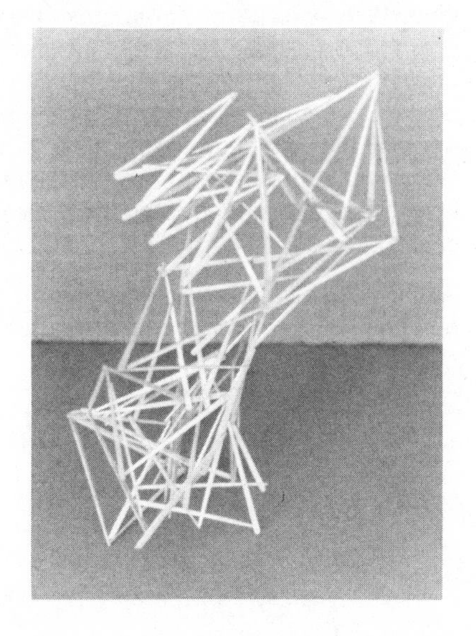

Paintings with Feeling

"Feeling" in a painting can be an emotional feeling, or the feeling you get when you touch the painted surface, or both at one time. In this project, the emphasis is on the tactile (touching) sensation one would get through the finger tips; but the finished product will appeal to the eyes as well as the fingers. It is one that says "please touch" rather than "don't touch."

MATERIALS

fine sawdust
white glue
heavy cardboard
newspapers
tempera paint
brushes (bristle)

PROCEDURE

A *heavy* cardboard painting surface is important because the amount of water used will cause warpage on thinner cardboard. The sawdust should be reasonably fine with any of the larger pieces removed.

Begin by brushing the white glue on the cardboard. If the cardboard is very large, do a small area at a time (about 6″ square). Sprinkle the sawdust onto the wet glue. With a piece of newspaper over the sawdust, press to ensure good bonding between the cardboard and sawdust. Remove the newspaper and tip the cardboard on its edge, shaking off the loose particles onto another piece of newspaper (this sawdust can be reapplied to another area). Repeat the process until the whole board is covered with a highly textured sawdust coating.

Apply a 50% water/glue mixture to the whole surface and then allow it to dry completely (overnight is best).

Now you are ready to paint. Paint with tempera that has a little white glue mixed in with it. If more texture is desired in some areas, simply glue on more sawdust or add it to the paint mixture. Paint freely to avoid very fine details, which will be difficult to achieve on this textured surface. A group of free-flowing flowers makes a good subject for this painting project, and will help to loosen up the child who tends to paint very tightly. Needless to say, the subjects are endless and not limited to flowers.

Woodcuts

One of the oldest forms of printmaking is the woodcut or wood block print. Any piece of soft wood scrap may be carved to make a printing block to be used to print designs on wrapping paper, fabric, or greeting cards. Woodcuts are often made in "limited editions" of five to 50 prints, each one being an original and signed by the artist. They are usually numbered with two numbers separated by a slash (1/25). The first number indicates the number of the print (1) and the second number shows the number of prints in the edition (25).

MATERIALS

soft wood scraps (pine or redwood)
wood cutting tools
printing ink
brayer (rubber ink roller)
newspapers
tray or glass for rolling out the ink
paper or fabric for printing

PROCEDURE

Ideas may be worked out on sketch paper before beginning on the wood block. A very important thing to keep in mind with printmaking is that the image will be printed in reverse, so any letters or numbers will have to be made backwards on the block.

When a satisfactory idea is established, draw it on the wood using a felt pen. Fill in the areas to be printed with ink. Since this is a relief print, only the raised part will be printed. All of the uncolored spaces will be cut away. *Caution*: The blades and gouges for cutting are very sharp, so use extreme care while working. The safest thing to do is to fasten the block to the table edge with a "C" clamp. Then, cut *away* from your body, simply turning the block when it is more convenient.

Cut *with* the grain of the wood whenever possible in order to make the tools easier to control. If there are any knots in the wood, it is wise to make them a part of the design rather than attempting to cut them out. Inexpensive wood-cutting tools can be purchased at supply stores. Some of these tools come with one handle and changeable blades. Others come in complete sets of several tools.

When all of the uncolored areas of the design have been cut away, the block is ready for printing. Spread out several layers of newspaper. Squeeze about an inch of ink onto a glass or plastic tray. Roll it out with the brayer, rolling in all directions until the rubber is evenly covered with ink. The object of the rolling is to cover the brayer evenly, not the entire glass or tray. So keep the ink in as small a space as possible.

Now roll the inked brayer onto the wood to cover the raised surface evenly with ink.

- If you are printing a fabric, be sure to use a permanent ink. Lay the fabric on newspapers for a padding. Place the inked block on the fabric and press evenly, with as much pressure as possible, over the entire surface. Standing on the block makes a good pressure—no shoes please!

- If you are printing paper for cards, wrapping, or a book, a water-base ink may be used. Place the *paper on the block* and rub the back of the paper with your thumb or the back side of a spoon. Be sure to rub all the way to the edges of the block to create a sharp, clear print.

In both cases, re-ink the block for each print made.

Texture by the Pound

Texture is an element that we see as well as feel with our fingers. This project will be an experience that is both visual and tactile, and will develop a clearer understanding of the concept of texture.

MATERIALS

wood scraps (soft wood; e.g., pine or redwood)
hammer
variety of metal pieces—nails, screws, bolts, nuts, chain, etc.
printing ink
brayer (rubber roller)
newspapers
paper for printing
glass or tray

PROCEDURE

This requires pounding and the noise can be close to deafening, so be prepared. No more than a few children should be pounding at one time. The object is to create a textured surface in the wood scrap, thereby creating a relief printing block. Any metal objects may be used. Lay the object on the block of wood and pound it with the hammer to make an indentation in the surface. Move the object and repeat the procedure, or remove it and do the same thing with a different form. Working on the classroom floor (or even better, outside) will reduce the sound. The children may choose to pound out a nonobjective design or one representing things such as imaginary birds.

When a satisfactory design is achieved, the block is ready for printing. Make a single print to be matted and hung or a repeated surface design.

Printing. Cover the work surface with newspapers. Squeeze about an inch of printing ink onto a glass or plastic tray. Roll it out in all directions using a rubber brayer. Roll it until the rubber is evenly covered with ink. The object is to cover the brayer evenly, not the glass or tray. Now roll the inked brayer onto the textured wood surface. Roll until all raised areas have been inked. Place the paper to be printed *on* the block of wood and rub the back of the paper with your thumb or a spoon. This will produce a sharper print than pressing the block on the paper. When completely rubbed, pull the paper off the block. Have the pupil re-ink the block for each print he or she makes. Sign it and hang it on the most suitable wall.

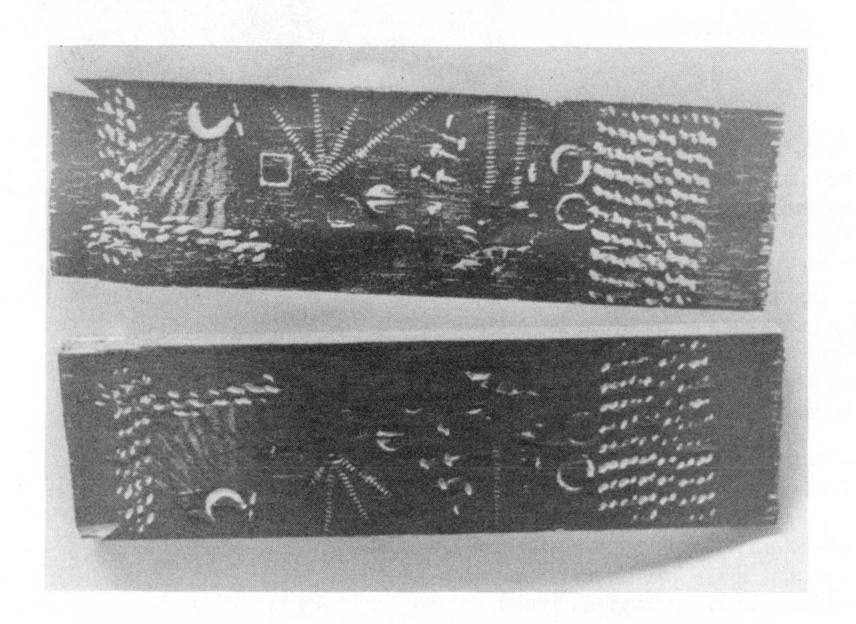

Spool Prints

Attention all people who sew—save those spools! Wood or plastic thread spools have another task to perform before you "retire" them. You can carve the round ends into a tool for making prints on paper or fabric. The designs will be somewhat radial or circular, providing a good opportunity for the teacher to talk about radial balance, such as that found in some flowers, wheels, etc.

MATERIALS

thread spools
craft knife (e.g., X-acto)
paint
fine sandpaper
paper for printing (or fabric)
newspapers

PROCEDURE

Remove the end labels from the spools (they may need to be sanded a little to remove all the paper and glue). The design will be made by cutting chips from the edges of each of the spools. Lines can also be cut into the flat surfaces of the ends. Two different shapes can be made from one spool by using both ends. Use more than one size spool to get greater variety in size while maintaining similarity in shape. *Caution*: The blades are very sharp, so work slowly and carefully. Haste may mean a cut finger.

Place the paper to be printed on a pad of newspapers for a cushioned printing surface. Apply paint to the spool with a brush, rather than dipping the spool in the paint. Dipping will fill the grooves and cause you to lose clarity of detail in the print. Lines may be added to the design with the edges of cardboard dipped in paint.

Your finished painting can be matted and hung or used as a folder cover. An old T-shirt might even be given new life with a printed design. Be sure to use textile paint in the latter case.

Cartons & Boxes

Who Said a Drum Must Be Round?

A square drum? Why not! If a square drum will make a sound, there is no good reason why we can't make it that shape. It can be played with the fingers or with drumsticks. Children of many ages can enjoy the rhythm of a square drum as it is tapped with sticks or fingers.

MATERIALS

square box (gift boxes are good)
white glue
scrap materials (e.g., yarn and fabric)
paint (water-base wall paint)

PROCEDURE

Glue the lid on the box. Simply apply glue to the ridge on the inside of the lid and put it on the box. For a little extra sound, put a few beans inside the box before gluing on the lid.

Give the whole box a base coat of water-base wall paint. There is usually a little paint left after a wall painting job. The water base solves the cleanup problem as well as provides a good base for glue-on decoration. A variety of scraps will enrich the sides of the drum. Keep the top plain or paint a design on it. A permanent paint should be used since tempera would be likely to flake off from the tapping.

Drumsticks can be made from cardboard rolls taken from pant hangers, dowel sticks, or tree branches. Even fingers will be effective rhythm makers.

Television Producers

Everyone is very much aware of the television viewing screen, so why not use it as the format for a learning experience for children? They can become involved in the production of a simulated TV show, from idea to script to visual layout.

MATERIALS

cardboard box
wood dowels or broom handle
shelf paper or wrapping paper
craft knife
paint or felt pens

PROCEDURE

Cut an opening in the side of a cardboard box the size and shape you wish your TV set to be. Make it slightly smaller than the width of your paper. Paint the box any color and add dials and knobs to suit your fancy. These could be drawn or made from cardboard shapes, buttons, or real knobs.

The scenes of the TV show will be drawn on the shelf paper or wrapping paper. This may have to be cut to fit the box your set is made from (it will have to be a little wider than the screen opening). The length of the paper will depend upon the number of scenes necessary for the presentation.

A script may be written by the children, or they may use a published story or play. Determine the scenes to follow the ideas of the script. Keep it rather short, perhaps no more than ten or fifteen scenes. Mark the spaces for each scene to fit the screen. Leave several blank frames for the lead and the ending. Then, with paint, felt pens, or crayons, draw the individual scenes in as much detail as desired. The first one would be the title, writers, producers, etc. One scene could run into the next or a sharp break may be indicated. Try several approaches for variety.

Now cut the dowels or a broom handle into two pieces about 2″ longer than the height of the box. (If the picture will move from bottom to top, as in the example shown, the dowels will go horizontally. If moving from left to right, the dowels will go vertically.)

Cut holes close to the front of the box for the dowels to slide through. Tape each end of the paper strip to the dowel (it is easier to put the dowels through the holes *before* fastening the paper to them). Roll the paper onto the dowel tightly. Be sure you have plenty of blank lead.

As the script is read, turn the dowels so the pictures change to fit the discription or dialogue. Play a record for background music if you wish.

Prepare to receive your Emmy Awards!

Chicken Box Crayon Box

The Colonel's chicken box (or anyone else's) makes a good storehouse for broken crayons or other art supplies. Broken crayons are often the best kind to draw and color with, but they usually need a new home as the supply of small pieces builds up. The decorated box will be easily identifiable, and if there are many boxes to be stored together they stack well. After pupils have enjoyed the "finger lickin' " goodies at home, have them bring you the box for recycling. (Save the bones, too, for jewelry projects!)

MATERIALS

chicken box
white glue
paint (tempera or acrylic)
felt pens with permanent ink

PROCEDURE

Paint the box with the acrylic paint, wall paint, or tempera paint mixed with a little white glue. Acrylic paint is probably best due to the use of the box, but ordinary wall paint will do equally well and there is usually a little left over after a big painting job. Second best is the white glue and tempera mixture.

When the paint is completely dry, decorate the box with felt pens. Again, due to the wear the box will receive, the permanent ink is necessary for durability. Each child will be able to identify his or her own box by the design, but a space may also be provided for a name. The design could be anything from butterflies to baseballs. In any case, it will be an individual solution to the problem of decorating a storage box.

Build and Rebuild

Building a large space construction can be fun. And even more fun when it can be done over and over again. The interlocking cardboard shapes in this project can be used many times, like a giant erector set. Towers, space ships, or whatever the imagination will bring forth are the fruits of this labor.

MATERIALS

corrugated cardboard (from boxes)
craft knife or scissors

PROCEDURE

Cut the sides out of your collection of boxes so you will have a series of flat shapes. Then, cut large geometric forms from these pieces. A craft knife is easiest to use, but extreme care should be taken when using the very sharp blades.

Now, cut slots into the sides of each cardboard shape. More than one slot can be cut into the longer sides. These slots should be narrower than the thickness of the cardboard to ensure a snug fit and a strong construction (see illustration).

When you have cut out a good supply of shapes, the actual construction begins. Simply slide the pieces slot to slot, considering balance and design as well as strength. Keep moving around the structure in order to develop it from all sides. Remember, three-dimensional construction in the round means that it is seen from all sides and the top.

Enjoy the completed design for a while—destroy it —and start again!

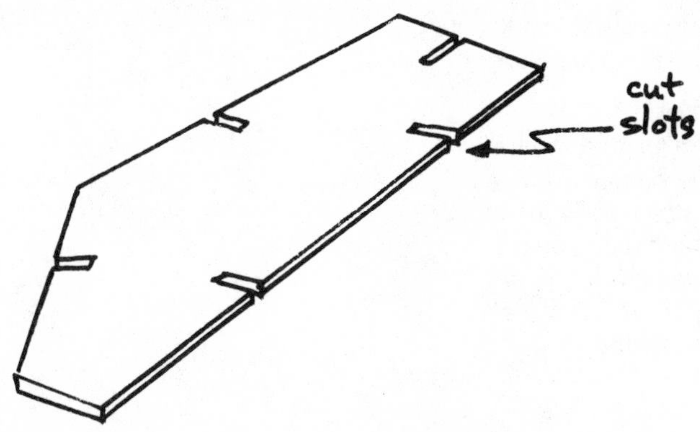

Weaving Around

Over and under, around and around may sound a bit like a carousel or leapfrog, but actually this is the weaving game. A woven purse or storage box is the result of this exciting box project.

MATERIALS
oatmeal box (or other round box)
craft knife
twine or carpet warp
blunt tapestry needle
white glue
yarn

PROCEDURE

Cut slits around the top edge of the box about ½″ apart and ¼″ long (see illustration).

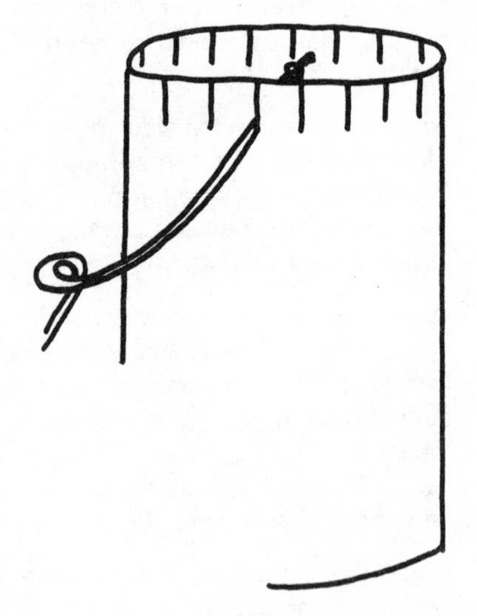

Knot the end of the twine or carpet warp and slide it into one of the slots with the knot on the inside of the box (see illustration). Count the number of slots and divide by 2 so you'll be able to find the center slot. Take the twine down the side, across the bottom, and up to the center slot on the opposite side. Now, take it to the next slot to the left and back down the side, across the bottom, and up to the slot on the left of the first piece of twine. Repeat this process until all slots are filled. Knot the last one on the inside.

Cut small slits around the bottom edge of the box for each string to slip through. This will hold them while the weaving is being done. These strings are called the warp, which is now ready for the weaving of the weft threads.

Weaving. A large-eye, blunt tapestry needle works best for weaving, but substitutions can be made, such as an open paper clip.

Thread the needle with about two arm lengths of yarn. Begin over-and-under weaving at the bottom edge of the box, working all the way around. When you run out of yarn, simply tuck the end under the threads next to the box. Continue with a new yarn length. If you are using scrap ends of yarn, random color selections can be made or a striped pattern can be created.

As you weave, pack each row against the previous one. Also, keep the warp strings (the up-and-down ones) straight and an even distance apart. You might be wise to check this each time you go around with the yarn.

Box Top. A top design can be made by simply applying yarn with white glue. A circle design or other shapes can be created. The glued yarn can be carried around the edge of the lid or a loosely braided strip might be applied. There's a lot of room for fun and variation.

A handle might be made with braided yarn and fastened to the sides of the box. Carry it or store it, but round and round we go to make it!

Tall Totems

A totem is a symbol for a tribe or family. Many American Indian tribes use totem poles, often with birds, animals, or other natural objects on them. Our box totems can be made as tall as ambition will take them and they can represent the group or individual making them. It might be good motivation to look at some pictures of Indian totems. However, explain to the pupils that copying designs is unfair and illegal, so they should only use pictures for ideas and inspiration.

MATERIALS

cardboard boxes of all kinds and sizes
paper
white glue
paint
masking tape
felt pens

PROCEDURE

Tape the boxes on top of one another at different angles. Size is almost unlimited. Very small ones, such as gift boxes, could be used, or very large ones, limited only by the door to get them in.

Draw mask designs with pencil or felt pens on the boxes. This will give some direction when the painting begins. Paint with tempera or acrylic paint. (If using tempera paint, the addition of some liquid soap will make the paint stick to the glossy surface found on some boxes.)

Any protruding parts, such as wings, ears, or beaks, may be made from cardboard or construction paper and glued or taped to the basic box shapes.

Lines indicating details should be the final touch after the paint has dried. The felt pens are very useful for this purpose.

Milk Carton Architects

Who designs the buildings in your city? Architects. And they have some restrictions and limitations put on their designs, just as you will have in this project. The size and shape of the cartons will create certain problems to work with. A whole city of buildings can be developed with a collection of cartons of all sizes and shapes, from the half pint container to the full gallon.

MATERIALS

milk cartons (all sizes)
tempera paint (with liquid soap)
scrap paper
white glue
masking tape
felt pens

PROCEDURE

Paint the cartons a basic color. Add a small amount of liquid soap to the tempera paint so it will adhere to the wax or plastic coating on the cartons. If acrylic or wall paints are used the soap is unnecessary.

Different shapes for the buildings can be made by taping several cartons together before painting them. Perhaps a skyscraper or two might even develop! Cut paper scraps to make windows, awnings, railings, etc. The children can add finishing details to the *dry* paint surface with felt pens.

Ridges and Rills

You and your pupils can feel *this design of high and low ridges and then see the light and dark patterns created by them. In other words, texture can be seen as well as felt, and this project will make this a very real concept to your class.*

MATERIALS

corrugated cardboard box
craft knife

PROCEDURE

Cut a side out of a cardboard box. Be sure that the box is made of corrugated cardboard or the process won't work. Any size piece will do depending upon your ambition. A light sketch with pencil may be made before the actual cutting begins.

Simple, solid shapes will be the easiest to cut, so keep this in mind as the planning proceeds. A sharp craft knife is a necessary tool for the success of this project. Use caution with the very sharp blade on this knife. It is wise to try cutting on a practice piece of cardboard before beginning on the final piece. Cut only through the top layer of cardboard, then peel it off to expose the corrugated center. It may be necessary to scrape part of the center to get the glue off the ridges.

Continue to cut and peel until the design is completed.

The natural color of the cardboard is pleasant, but it could be painted another color if desired. This could also be used as a printing block for relief printing. Just coat the raised surfaces with paint or printing ink and transfer the design to paper.

Easter Beauties

Plastic eggs originally hold stockings or pantyhose, but after they have been emptied children can use them for many projects, including Easter decorating. Send out the word to save pantyhose eggs!

MATERIALS

plastic eggs
felt pens with permanent ink
household cement
scrap trimmings from sewing projects

PROCEDURE

The closed egg will provide a surface for many imaginative, decorative designs for Easter. Perhaps your pupils will draw designs on the eggs with felt pens. Or they can glue on a wide variety of materials to create a surface design (use quick-drying household cement).

The open egg might be used to create an enclosed scene. The possibilities are endless. Just start with the egg and your imagination.

At Christmas time the eggs could become tree ornaments using the same procedures but with a different intent. Simply glue on a loop of string for a hanger and it is ready for the tree.

Still another idea is to use the egg as a gift box for those very small gifts. They'll never guess what's inside of this one!

Dessert Box Talkers

Small boxes containing pudding, gelatin like Jell-O, or other desserts, or individual cereal boxes should be saved after the contents are used. The boxes can be made into "talkers"—wonderfully simple puppets. This is the kind of puppet that does not need a body to be effective, just the head.

MATERIALS

dessert boxes or individual cereal boxes
masking tape
tempera paint
white glue
felt pens
craft knife
variety of scrap materials (e.g., yarn, fur, etc.)

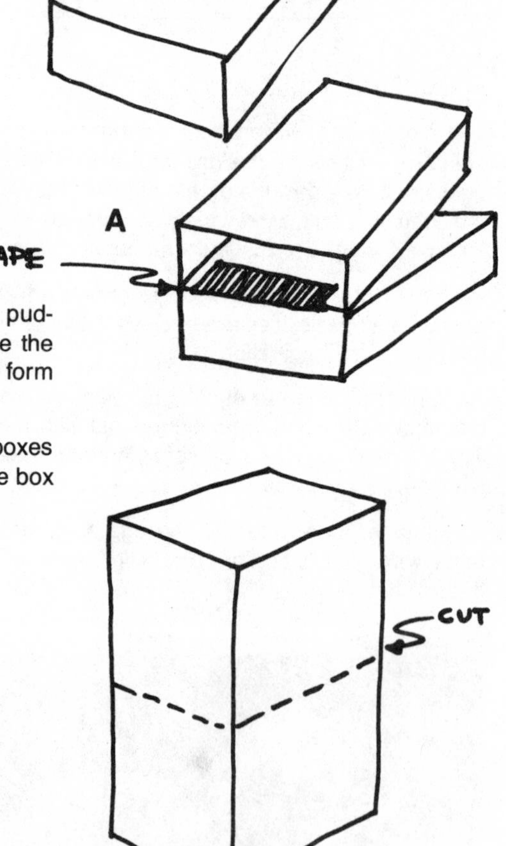

PROCEDURE

Dessert box. Cut one end out of two gelatin or pudding boxes. Put one on top of the other and tape the open ends together using masking tape. This will form the jaws of the talker (see Illustration A).

Cereal Box. These are the small, individual boxes that come in a variety pack. Cut three sides of the box around the middle (see Illustration B).

Fold it back on the uncut side. A piece of masking tape along this fold will help to strengthen the talker's jaw (see Illustration C).

Both techniques proceed in the same way from this point. Develop the character by adding teeth, tongue, eyes, ears, hair, etc., with any number of materials you might find.

To make the talker "talk," place your thumb in the bottom box and your fingers in the top box. He's ready to tell a story, give a report, or participate in a play.

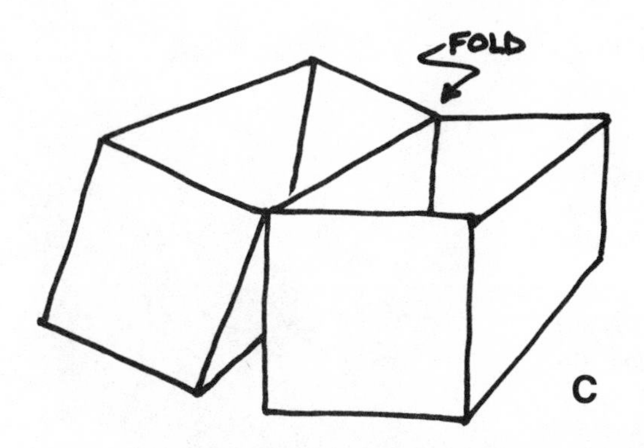

Relief in White

Value, or the light and dark element in design, is dramatically demonstrated in this cut cardboard relief. The chipboard used in many gift boxes or the backing on shirts and writing pads is a good supply source for your material.

MATERIALS

cardboard (not corrugated)
scissors
white glue
white tempera paint
brushes

PROCEDURE

Since this is a relief design, the shapes used will be raised slightly from the background, thus creating the light and dark patterns. The largest shapes should be cut and glued on first. Then, gradually, reduce the size of the added pieces.

Cover the entire back of each piece with a light coating of white glue to make sure it lies flat as it dries. By repeating shapes in different sizes, you will help to create a unified design.

When several layers have been added to complete the design, allow it to dry completely. Then paint it with white tempera paint. More than one coat may be necessary to get an even coverage of white. The result of light striking the surface will be a dramatic light and dark pattern. Try a very strong light source from one side or the other to make darker shadows. Play with it. Enjoy it. Learn from it. Light and dark patterns can be a great deal of fun.

Chipboard Constructions

Flat geometric planes cut from chipboard combine to make space constructions that can be very simple or very complex. An increased awareness of what we mean by "large" and "small" will develop from the problems involved in assembling the shapes.

MATERIALS

chipboard from old gift boxes or writing pads
white glue
scissors
paint (optional)

PROCEDURE

Cut simple, straight-edge geometric shapes from the cardboard scraps. These can be from 1″ to 6″ in size.

Apply white glue to the edge of the chipboard shape and place it on a flat side of another shape. Hold it a moment until the glue sticks and dries a little.

Continue adding shapes, changing directions as you build. Use some pieces as supports as well as design elements. A variety of sizes will help make a successful construction.

To add some emphasis to certain parts, you might paint some flat surfaces white to contrast with the gray cardboard.

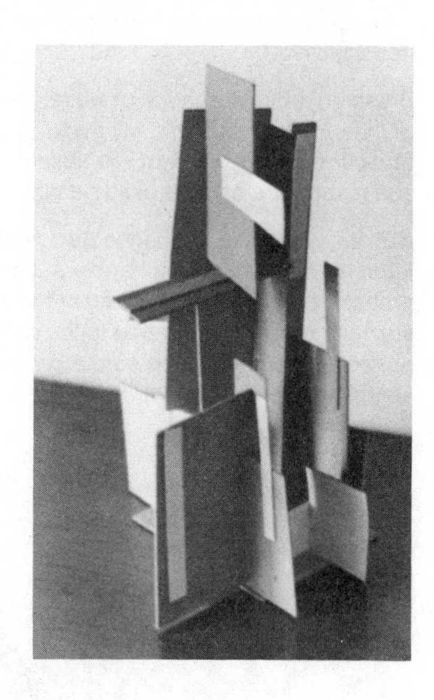

Cottage Cheese Tub Planters

The familiar cottage cheese tub makes a fine planter or planter holder. The technique of "yarn painting" used by the Huichol Indians of Mexico makes a beautiful covering for such a tub. When decorated, the tub makes a delightful addition to the decor of a kitchen, bedroom, or bath. And it's a great activity for using up yarn ends.

MATERIALS

cottage cheese tub
colored yarn scraps
white glue

PROCEDURE

The design can be a simple representation or purely nonobjective.

Brush on a heavy coating of white glue to a small area of the tub at a time. This will avoid getting excess glue on your fingers. The yarn will absorb the glue quickly, so be sure you apply a heavy coat.

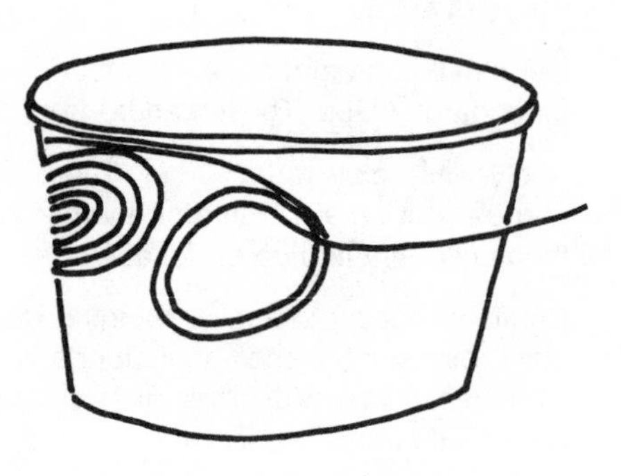

Cut the yarn into manageable lengths. You'll soon learn how much you can handle at one time, which will probably be about 24". Starting at the outside edge of a shape and working your way to the center is usually the easiest, but not a rule. Be sure all ends are glued down securely (see illustration).

Berry Basket Easter Basket

Easter baskets or May Day baskets can be made from the plastic tomato or berry baskets found in the market. The weaving experience is good for younger and older children alike.

MATERIALS

plastic berry basket
yarn or ribbon scraps
construction paper
white glue

PROCEDURE

Collect scraps of colored yarn or gift package ribbon.

Begin weaving in and out of the vertical strips on the sides of the basket. Alternate over and under with each successive trip around the basket (see illustration).

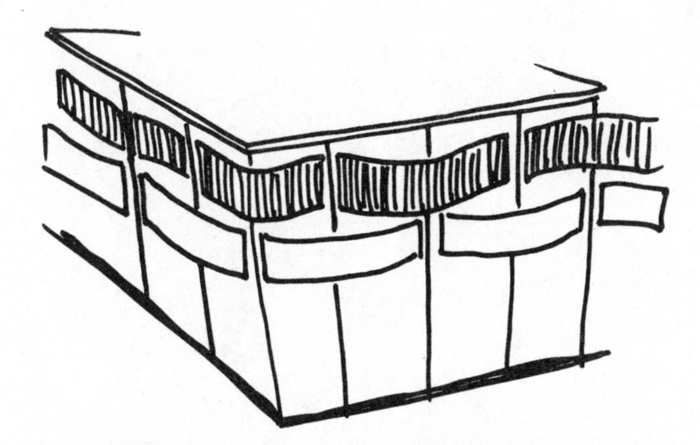

Cut a strip of construction paper ½″ to 1″ wide, and long enough to make a handle. Glue it into opposite corners of the basket. This will be a purely decorative handle. Don't try to lift the basket with it.

Fill the basket with goodies for Easter or May Day, or maybe set a small potted plant in it.

Cans & Pans

Sculptures That Sparkle

Mobiles or stabiles, these sculptures that sparkle are no more than a blend of aluminum pie pans and imagination. The sparkling objects that result have many uses and can become ornaments for a tree at Christmas. So ask your pupils to begin saving the pans from frozen or bakery pies and even the pans from those time-saving potpies.

MATERIALS

aluminum pie pans
scissors
wood block for base (optional)
brass fasteners
stapler
permanent ink felt markers (optional)

PROCEDURE

Free forms can simply grow and develop as you cut into the metal pans. Cut and fold! Bend and twist! Punch with a paper hole puncher! Let yourself go! Discover what happens when the light reflects off of the smooth surface. Try texturizing some of the sculpture with a sharp tool. Move slowly to avoid cutting your fingers. A little caution can completely eliminate cuts and scratches.

A stapler will be handy for mounting the sculpture on a wooden base. Small tacks will also do the trick. Brass fasteners will hold pieces together when you are overlapping them.

For a mobile (a sculpture that moves), suspend your sculpture from a string or nylon fishing line. One or several shapes can be very effectively hung in the corner of a room or from the light fixtures. The movement of air in the room will make your sparkling sculpture turn gracefully.

If a touch of color is desired, use permanent ink felt markers on one side of the shapes to create a contrast with the pure metal. This is particularly effective for Christmas ornaments, but leave some of the metal in its natural state.

Cans of Ivy

These cans of ivy made from coffee cans might be planned for a Mother's Day or Father's Day remembrance, which will continue to say "love" all year. The planting and growing could very well be part of a science lesson.

Start collecting one-pound coffee cans well in advance of the three or four days necessary for the completion of this project.

MATERIALS

1 lb. coffee can
white glue
bristle brushes
scissors
construction paper (from scrap box)
newspapers
breakfast cereal or macaroni (optional)
shellac
tempera paint

PROCEDURE

Cover your work surface with newspaper to protect it from glue. Cut designs from construction paper or any similar paper. Color is unimportant, since the paper will be painted anyway. When planning the shapes of the design, keep in mind that small spaces between shapes will add to the finished appearance. This can be accomplished by cutting and sliding pieces apart or by putting one piece on top of another (see illustration).

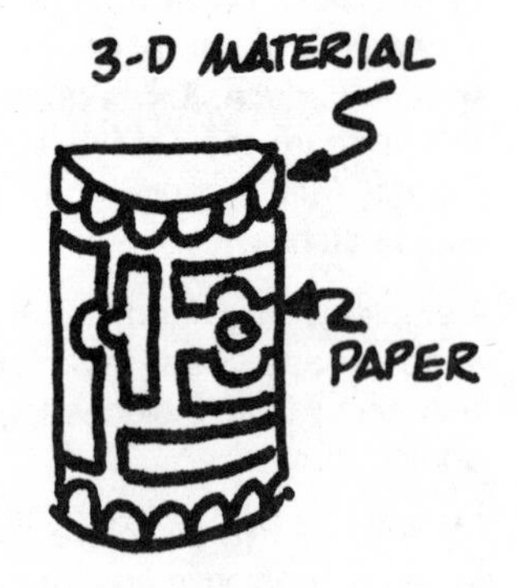

Apply a coating of white glue to the surface of the can before placing the paper shape on it. Then, brush *over* the paper with more glue. Keep brushing until the paper sticks and conforms to the shape of the can.

Three-dimensional materials may be glued to the top and bottom edges for additional interest. Such things as breakfast cereal with regular shapes, dried peas, or macaroni in its various forms can be used. Coat these materials well with glue after they are secured to the can. When the glue dries it will protect them, as well as make them hard. Allow to dry overnight.

Paint the can black, brown, or other dark color, being certain to get into all the grooves and edges. A little white glue mixed in with the paint is helpful. Allow to dry completely, preferably overnight, before the next step.

Next, paint over the dark color with a light color, using a *very dry* brush. Dip your brush into the paint and wipe most of the paint off on a piece of newspaper. Then, quickly brush across the surface, avoiding too much color at one time. By putting on a little at a time in several directions you will create an antique effect. Scrubbing too hard will tend to loosen the dark color already dried on, so work gently. Dry again.

Finally, brush on a coating of shellac or clear acrylic medium to provide a protective finish. Apply very gently to avoid disturbing the paint. You may use a spray finish coating if you have proper ventilation or facilities to work outdoors. Avoid using a spray in a classroom though. Spray fumes can be toxic and the overspray can certainly make a mess.

Your container is now ready for planting!

Masks with a Shine

Your pupils will enjoy making these masks. They'll need one or two aluminum pie pans to create a decorative wall mask with a shine. The tin masks of Mexico would be good motivation, starting the imaginative transformation of the aluminum pan. This project is recommended for older children because of the dexterity required.

MATERIALS

aluminum pie pans (9″ or larger and 4″ potpie size)
scissors
brass fasteners (size #00)
sharp nail
larger brass fasteners for decoration (optional)

PROCEDURE

The 9″ size pan is used as the base for the mask, since smaller sizes would be very difficult to handle. Think of the possible facial features that might be included on the mask: bulging eyes, raised eyebrows, eyelashes that curl, big ears, mustache or beard, elaborate headdress, etc.

When all this is determined, the problem begins. If embossed designs are planned, it is best to do them from the back side (the inside of the pan). A ball point pen or sharpened stick will work well for this technique. Place the pan on a stack of newspapers for a cushioned surface and use firm, even pressure for the best results.

Cut the shapes for added features (eyes, nose, etc.) from other pans of any size. The small potpie size is fine for this part of the mask. If you are unsure about cutting the shapes directly, they may be cut from scrap paper first and used as a guide in cutting the pan.

To fasten the shapes to the base of the mask, use the small #00 brass fasteners (see illustration). Make small nail holes in the pieces to be joined by simply pushing the nail through them. Also make small nail holes in the appropriate places on the base of the mask where the added pieces are to go. Flatten an edge on these pieces so they will fit more easily.

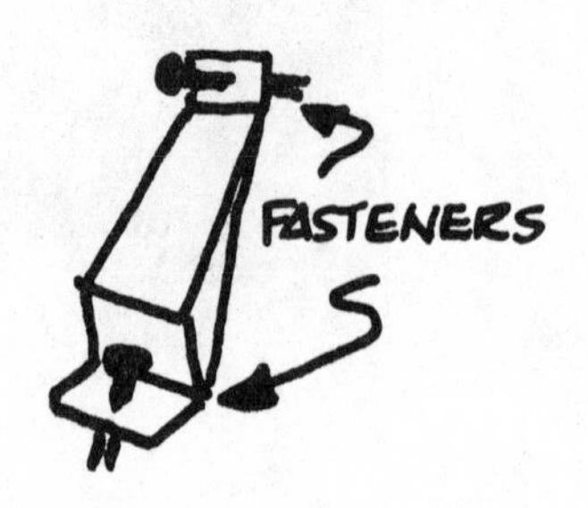

Larger size brass fasteners can be used for additional surface enrichment of the mask. Foil wrap may also be used for some details if the stiff metal of the pan is too difficult to manage. The foil wrap is easily crushed and shaped to any form, but doesn't give quite the same effect as the pans do.

A ring pull from a soft drink can taped to the back of the mask serves as an effective hanger for mounting your "mask with a shine" on the wall.

Pencil Stower

The Huichol Indians of Mexico often use "yarn painting," a craft which consists of pressing brightly colored yarn into soft beeswax. Odds and ends of colorful yarns can transform a simple container into a beautiful and functional pencil stower. Your pupils can decorate these containers easily, and the result will be an attractive stower which helps keep pencils, pens, and brushes conveniently at hand.

MATERIALS

frozen juice can
bristle brush
yarn scraps
newspapers
white glue
scissors

PROCEDURE

A large or small frozen juice can is particularly suitable for this project because most of them are made of heavy cardboard with a metal bottom. However, a vegetable can will do if juice cans are not available.

The design for the pencil stower could be either a simple representational design or a purely nonobjective one. A limited color scheme of one dark, one medium, and one light color is most effective. You can have your pupils draw on the can first or they can develop their design freely as they work.

In either case, brush a heavy coating of white glue onto a small area of the can at a time. This will avoid getting excess glue on your fingers and also prevent drying before you get to all of the parts. A damp sponge close at hand will help keep the yarn from sticking as you work. If you pour a small amount of glue in a shallow container, such as a pie pan or cottage cheese tub, you'll find it much easier than using glue out of the squeeze bottle. The yarn will absorb the glue quickly, so be sure to use a heavy coating. You'll also find that cotton or wool yarns are easier to work with than acrylics.

Cut the yarn into manageable lengths, since the joints of the same color won't show. You'll soon learn how much you can handle at one time, which will probably be about 24″ maximum. Starting at the outside edge of a shape and working your way to the center is usually easiest, though not always so. Be sure all ends are glued down. Put the can on its side with the fingers of one hand inside it to give you more control (see illustration).

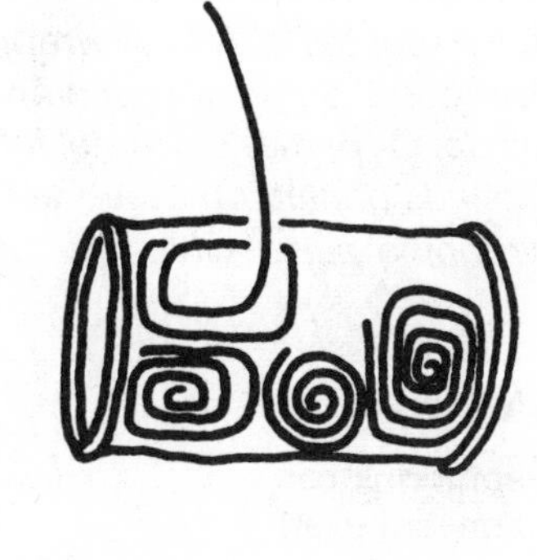

When it is dry, stow your pens, pencils, and brushes in a place where you'll always know you can find them.

Print-a-Can

Your student can turn an empty soft drink can into a tin can printer for making decorative paper for folders or wrapping. A large portfolio for keeping all of the artwork for the year could be printed with an all-over pattern created by the same artist whose work is inside. Or perhaps colorful folders could be made for science or social studies reports. Designs printed on tissue paper, butcher paper, or shelf-lining paper will make unique wrapping paper which will further individualize gifts.

MATERIALS

soft drink can
tempera paint
newspapers
white glue
heavy twine (16-ply or felt scraps)
scissors
brushes
construction paper
wrapping paper

PROCEDURE

Cut one end out of the can, preferably with an electric can opener because it leaves a smoother edge and will help to avoid cut fingers. If necessary, cover the cut edge with cloth tape as an extra safety measure. The printing design goes on the curved side of the can. This design can be made with the 16-ply twine if a line design is planned or put together with pieces of cut felt if solid shapes are desired. The two could be combined when printing, but not on the same can. The reason for this is that the twine and felt may be a different thickness, and therefore would not touch evenly when printing. The twine needs to be thick enough so it is higher than the ridges on the bottom and top of the can.

If you are using twine for a line design, glue the end of the twine to the can and overlap this spot as you wrap tightly around the can. Overlapping itself here and there will cause a space in the printed line, adding interest to the pattern. Glue the finishing end of the twine on another point on the can (see Illustration A).

If you are using felt, cut shapes from the felt scraps and arrange them on the table before gluing them to the can. Put glue in a shallow container, such as a pie pan.

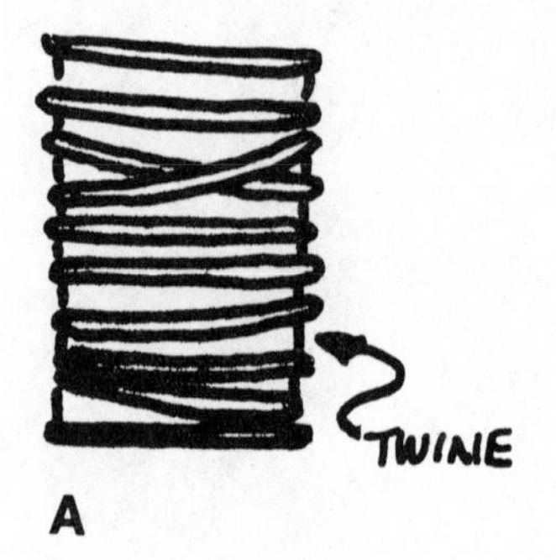

A

Saturate one side of each felt piece by dipping it into the pan of glue. Place it on the can, being careful that the whole piece is making contact with the can's surface (see Illustration B).

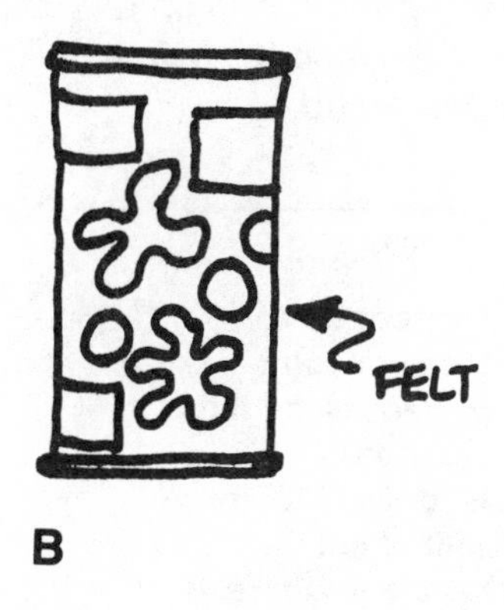

B

Dry thoroughly before printing. Overnight is best.

Thin the tempera paint enough so the twine or felt will soak it up and remain wet while printing is being done. Apply the paint to the twine or felt with a watercolor brush. Be sure to saturate well.

Place the paper to be printed on a stack of newspapers to make a cushioned surface. With two or three fingers of one hand inside the can for even pressure, and the other hand guiding the direction, roll the printer slowly across the paper. You should be able to roll the printer across twice before needing to apply more paint. Don't try to print too rapidly or the printer will slide and smear your design.

Pop-Up Puppet

Let a puppet pop up and do the talking! The personality created by facial expression and costume will help a child use his fanciful imagination to tell stories or make reports to the group.

MATERIALS

1 lb. coffee can
tempera paint
papier-mâché mix
brushes
½″ dowel (12″ long)
white glue
needle and thread
fabric (about ¼ yard)
miscellaneous objects for decorating (e.g., buttons, yarn, etc.)

PROCEDURE

Puppet Head. Model a papier-mâché head about the size of a golf ball on one end of a ½″ dowel which is about 12″ long, This papier-mâché mix can be purchased commercially or it can be made from shredded newspapers and wheat paste. Fingers, popsicle sticks, and pencils make good modeling tools. Leave a flange at the bottom of the neck for fastening the costume (see Illustration A). When completely dry, paint with tempera. Give the head a final coating of white glue or clear acrylic medium for protection.

Costume. About one-quarter of a yard of fabric of any kind is needed for the costume (this will also cover the coffee can home of the puppet).

Fold the fabric in half so the fold is at the top. About 6″ x 8″ will be needed to cover the can. Measure up about 6″ from the bottom of your piece of fabric and begin to taper in to the underarms of the puppet, then out to form the arms. Cut a neck hole just large enough for the dowel and flange of the neck to fit through (see Illustration B). Turn the fabric inside out and stitch up the sides, leaving about a quarter inch seam. (A sewing machine could be used for this.) Hands cut out of felt may be sewn into the arm holes before turning the fabric right side out.

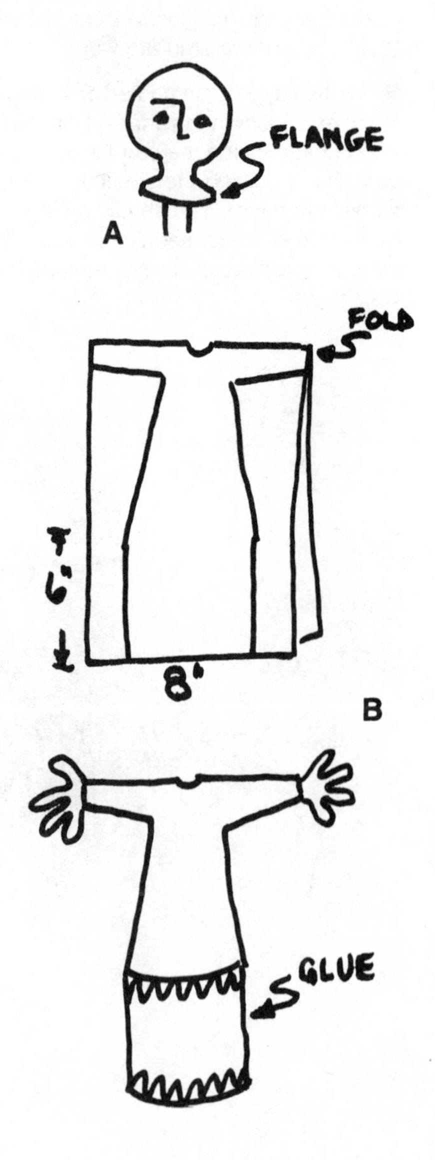

Slide the costume over the coffee can and glue the bottom section to the can. Place the painted head, stick first, into the neck opening. You may either glue the fabric to the neck or stitch it tightly. You are now ready to add objects for enrichment, such as hair made of yarn, buttons, fur, lace, etc.

Papier-Mâché Recipe. Tear newspaper into small pieces (no larger than 1″ square). Put them into a bowl or bucket and cover with hot water. The quantity will go down, so be generous. Let it stand overnight.

Squeeze out all the water you can. Work it by pulling apart with your fingers to make it mushy. Add diluted glue (about half water). Mix with your hands until it all sticks together. It will look and feel like clay and can be shaped easily.

Pans with a Beat

Everyone likes to get into the act when the music starts to play, and what better way to do it than with a beautiful rhythm instrument? This is a good opportunity to integrate an art project into the music program using shiny aluminum pie pans or colorful paper plates. Each of these will produce a different kind of sound.

MATERIALS

2 aluminum pie pans (or 2 paper plates)
beans, peas, or macaroni
permanent ink felt markers
or tempera paint
brushes
paper hole puncher
yarn and tapestry needle

PROCEDURE

The aluminum pie pans can be decorated with permanent ink felt markers; the paper plates with felt markers or tempera paint with a little white glue mixed into it for permanence. Decorate the pans on the bottom, which will become the outside of the instrument. If you decorate both of them before putting them together, you will be able to set them aside to dry at the same time. Paint them with flowers, dots, stripes, or an op-design.

Dry beans, peas, and macaroni serve as noisemakers inside the pans. Try several different things and vary the amounts of each that you put in.

Lacing the edges together is done simply with colored yarn. Choose yarn colors which coordinate with the colors used in the designs on the pans. A paper hole puncher will make lacing the aluminum pans easier, but a tapestry needle will work for the paper plates. Use a whip stitch, going down through each hole all the way around the pans (see illustration). Tie a bow, add a bead, or make a tassel or two. This will finish off the lacing. Then get the music started!

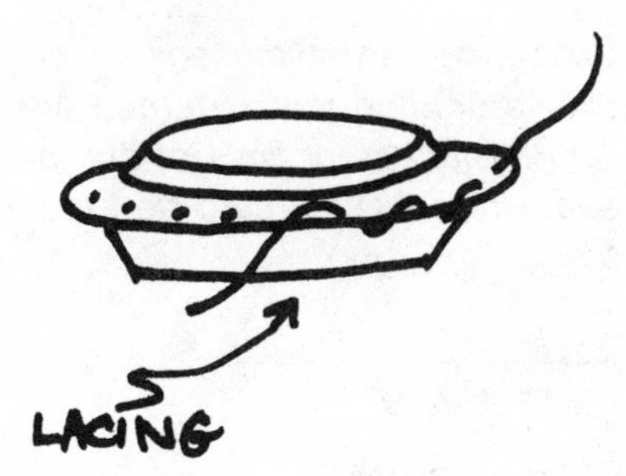

Fish Plate

Suspended from the ceiling or a light fixture, these graceful fish can transform a classroom into the undersea world of Jacques Cousteau. You might first show your pupils pictures or films of the many beautiful and exotic types of fish that swim in the sea, since these would be certain to stimulate young imaginations to create fish fantasies.

MATERIALS

2 paper plates
scissors
white glue
tempera paint
construction paper
brushes

PROCEDURE

First, cut a tail, fins, and a mouth from colored construction paper. The two paper plates placed on top of one another form the two sides of the body of the fish. Apply glue all the way around the edge of the paper plates. Position the construction paper parts on the glue and then place the second paper plate on top, forming a "fish sandwich." When the glue is dry, begin painting the body of the fish in beautiful colors and patterns. Felt markers could also be used instead of, or in combination with, the tempera paint.

Suspend your fish from the ceiling or light fixtures with a string or nylon fishing line. The currents of air will cause the fish to swim playfully in a gentle motion.

Mask on a Stick

The "real me" can come alive with a mask on a stick. A shy child may find it easier to talk to the group, to recite, or to tell a story from behind a false face. The characters of a skit or play could be designed using a set of stick masks, much as the ancient Greek actors used masks.

MATERIALS

¼" or ½" dowel (12" long) or cardboard tube from a hanger
tempera paint or felt markers
brushes
scissors
white glue
variety of scrap materials (e.g., yarn, buttons, etc.)
paper plate

PROCEDURE

Glue the dowel or cardboard roll from a clothes hanger onto the back of a paper plate. This will be the handle for holding the mask in front of the actor's face.

Paint features on the plate to create the personality of the character being created. Add, using white glue, any variety of materials for hair, eyes, beard, etc., to increase the mask's effectiveness. Simple, but fun to use. And when not in use they may be good as wall decorations.

Weeds Grow Tall

Children and adults often want to display dried leaves, weeds, or flowers collected on a nature walk. The entry hall, patio, porch, or hearth are good places for these, but often their stems are too long for the containers most readily available. Here is a project for a tall "weed holder" that could hold tall cut flowers as well.

MATERIALS

3 or more 2 lb. coffee cans
newspapers or paper towels
white glue
bristle brush
scissors
masking tape
tempera or acrylic paint
yarn

PROCEDURE

Remove the bottom from two of the three cans (if more are used, leave one bottom in). Use masking tape to secure the cans together, pulling the tape as taut as possible to ensure a snug fit.

Cover the entire surface of the cans with small pieces of newspaper or paper towels, spreading the glue evenly under *and* over each piece as it overlaps another. Allow this to dry before you continue to the next step (see Illustration A).

Then, dilute two parts of white glue to one part of water in a low, wide-mouth container, such as a cottage cheese tub. Cut the yarn into pieces about 24" long so they are easy to manage. Dip the yarn into the glue mixture and stir with a brush until the yarn is saturated. Hold the yarn against the side of the container with the bristle brush as you draw it out with the other hand. This will squeeze off excess glue but won't squeeze it dry. Drape the yarn, as though drawing on the can, in the desired pattern, guiding it with the brush as you work. Tap the yarn with the end of the bristles to ensure contact with the surface all the way along. When the design is completed, allow to dry about a day before continuing (see Illustration B).

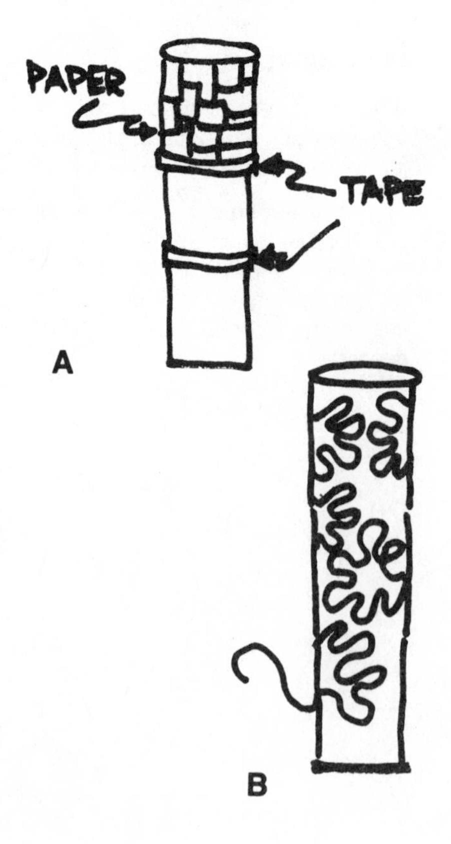

Paint the whole surface with a dark color. A little white glue mixed in with the tempera will make a better surface for the next step. Or, use acrylic paint. Dry thoroughly.

An antique effect, which emphasizes the line design of the yarn, is achieved by dry brushing with a light color or two. Dip a bristle brush into the light-colored paint and wipe most of it out on newspaper. Then, quickly brush across the surface avoiding too much color at one time. Do this in all directions several times until you achieve the color contrast you want. Allow to dry.

Fill the holder with your collection of dried weeds and sit back and enjoy.

Can Coasters

Many of your students will have pet cats at home, and the kitties will be able to help in this project by eating lots of cat food. Those who don't have cats can substitute other cans of similar size. Sets of four or six coasters will be welcome gifts and they are simply made, with plenty of room for individualization.

MATERIALS

cat food cans (4 to 6)
newspapers or paper towels
yarn or heavy twine
tempera or acrylic paint
white glue
enamel paint (optional)
brushes

PROCEDURE

Remove the labels from the cans. Paint the inside with enamel or acrylic paints if you object to the plain metal color. Tempera paint shouldn't be used on the inside of the can, since moisture from wet glasses set in the coasters will spoil tempera coatings.

Cover the outside of the cans with newspaper or paper towels torn into strips. Brush out the glue to completely cover the surfaces under *and* over the strips as they are applied. Allow to dry.

Dilute two parts of white glue with one part of water in a low, wide-mouth container. Cut your yarn into pieces no longer than 12". Dip each piece into the glue mixture and stir it with the brush until it is well saturated. As you draw the yarn out of the container with one hand, hold it against the side of the container with the bristles of the brush. This will squeeze out excess glue without making it completely dry. Now, drape the yarn as though you are drawing a design on the can, guiding it with the brush as you work. Tap the yarn with the end of the bristles to ensure contact with the can. Allow to dry before painting.

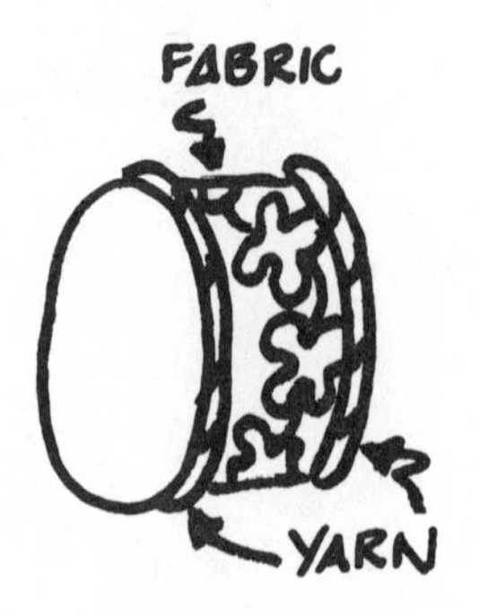

Paint the whole surface with a dark color. A little glue mixed into the tempera paint will make a better surface for the next step. Or you may use acrylic paint, which is more permanent. Dry thoroughly.

An antique effect, which emphasizes the line design of the yarn, is achieved by dry brushing with a light color or two. Dip a bristle brush into the light-colored paint and wipe most of it off on the newspapers. Then, quickly brush *across* the surface avoiding too much color at one time. Allow to dry completely.

A spray of lacquer or clear acrylic medium will give a protective coating to these delightful coasters.

Giant Blossoms

There may be some Texans who will want to claim these giant blossoms for their own, but they could be found anywhere and everywhere. As individual projects, they will make proud gardeners of your students. In a mural, they will make a beautiful, bountiful garden to enhance a spring bulletin board or an entire wall.

MATERIALS

paper plates
brushes
scissors
tempera paint
¼″ dowels (optional)
butcher paper (for mural)
liquid soap
construction paper

PROCEDURE

You may use paper plates that have been used and washed for these flowers. Cut sections out and glue them back on another place on the plate, or cut and fold shapes up to create unusual flowers.

Paint them with tempera paint that has a little liquid soap added to it (to make the paint adhere to the waxy coating found on some paper plates).

When the paint is dry, mount the flowers on a dowel and add construction paper leaves.

If you are making a mural, mount them on butcher paper in great numbers.

Another possibility is spring banners with flowers mounted on both sides of the butcher paper panels and suspended from the ceiling.

Dried Flower Wee Ones

Hanging in a wall grouping or standing on a desk, this colorful little accent will be sure to brighten the eye and lift the spirit. The dried flowers or weeds for these "wee ones" can be collected as a group or individual assignment. Perhaps some small rocks or seashells might also be included as accents in the arrangement.

MATERIALS

cat food can (or similar size can)
tempera paint or acrylic paint
scrap fabric
brushes
white glue
pull-top ring
yarn
masking tape
flower clay
dried flowers or weeds

PROCEDURE

Remove the label from the can and be sure it has been washed out. Paint the inside of the can with a color that will harmonize with the fabric and contrast with the flowers or weeds. If tempera paint is used, add a little white glue to make it stick to the metal. Allow to dry.

Cut a piece of fabric to fit around the outside of the can with about ½″ overlap. Cover the can with white glue, brushing it out evenly, and apply the piece of fabric. (Before it is glued to the can, the fabric could have a stitched design added to it.) Glue a ring of yarn around the top and bottom edges of the can to cover up the raw edge of the fabric (see photograph). Tape a ring pull from a drink can onto the back for a hanger. Securely fasten the flower clay to the inside of the can. Arrange the flowers or weeds and whatever else is being used, and then hang it on the wall and wait for the ohs and ahs!

Fabrics & Buttons & Bows

Get the Feel of It

These fabric collages are really tempting to the touch. Collage *is design created by arranging materials of various colors and textures on a surface. Your pupils will develop an awareness of the tactile quality of things as they organize the variety of fabrics collected. These collages can be permanent or changeable. After a design is organized, it may be changed by rearranging the shapes at another time.*

MATERIALS

fabrics of all kinds
scissors (pinking shears are useful)
white glue
box lid (if changeable collage)
heavy cardboard (if permanent collage)

PROCEDURE

Cut shapes from the gathered fabrics. Arrange them in the box lid or on the cardboard, emphasizing variety in textures (from smooth to rough) as well as variations in size from large to small shapes. Rearrange the shapes several times to try different looks and feelings. If it is to be a permanent collage, use white glue after the final arrangement is determined.

Just looking at these finished products will invite the fingers to explore the surface variations. The process and the product are tactile experiences.

Banner Festival

Banners for celebrating any kind of event, special group, or person can be designed from all sorts of different fabrics. Norman LaLiberti is probably one of the best-known banner designers. He uses all kinds of materials and tools to make his banners. If you can find any pictures of his work, they would provide good stimulation for a group embarking on this project.

MATERIALS

large fabric for the foundation of the banner
scrap fabrics
white glue
needle and thread
scissors (pinking shears helpful)

PROCEDURE

Iron the fabrics if they are wrinkled, because it makes them easier to work with. Help your pupils decide on a theme to be developed on the banner.

Cut out shapes (pinking shears help to prevent raveling edges) and lay them on the foundation fabric. Arrange and rearrange them until the most satisfactory design is achieved. When you are happy with the results, begin gluing or sewing the pieces in place.

Buttons, yarn, and any other materials might be used to enrich the surface.

The bottom edge of the banner may be finished by hemming, fringing, or adding tassels. Be as inventive about the bottom edge as with the design of the banner body itself.

Have a festival, fiesta, or some sort of celebration to display the banners.

Face It

A mask that you can see through and yet hide behind will give a shy child confidence in speaking. Your pupils will readily tell a story, give a report, or create a character with the help of a nylon stocking and a wire coat hanger. Halloween and Mardi gras are good times for this project, but it is not limited to those holidays.

MATERIALS

wire coat hanger
scrap fabrics
yarn
colored paper
nylon hose (one leg of pantyhose is okay)
white glue
scissors
needle and thread

PROCEDURE

Stretch the hanger by pulling on the cross bar and hook in opposite directions (see illustration). Slip the nylon stocking over the wire and tie it at the top and bottom. Decorate with features using fabrics, yarn, etc. White glue and needle and thread can be used to fasten eyes, nose, mouth, etc. to the stocking.

Tell your students to hold the finished masks in front of their faces and become different persons. Encourage them to let themselves go, because they are now completely different characters without inhibitions.

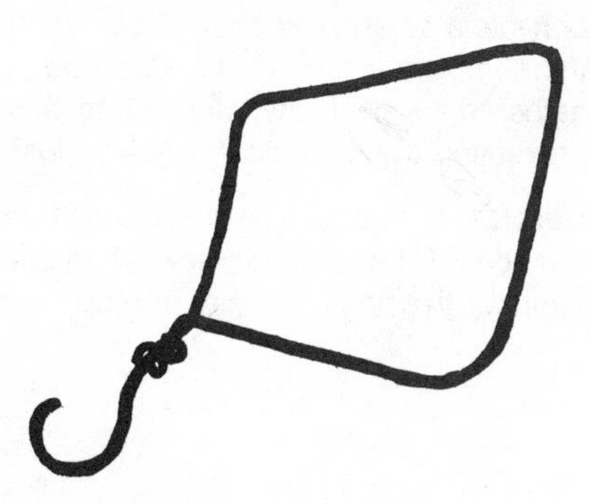

Lost Your Marbles?

The game of marbles goes back to the days of Egyptian and Roman children and is still a favorite in present-day America, where tournaments are held annually. Remind your class that marbles are easily lost if they are not kept in a bag. So why not have each member of the class make one, using his or her very own design?

MATERIALS

scrap fabric (at least 9″ x 12″)
needle and thread
heavy twine or ribbon
materials for optional decoration (e.g., felt scraps, felt pens, stitcher)

PROCEDURE

Fold a one-inch hem along a long edge of the fabric (with the *wrong* sides of the fabric together). Three-fourths of an inch from the fold, stitch with small, close stitches (see Illustration A).

Fold it in half with the *right* sides of the fabric together. Begin at the stitches of the previous hem and stitch down the long side and across the bottom. Stay ¼″ to ½″ from the edge of the fabric, using small, close stitches. (see Illustration B).

Turn it inside out. Loop a length of heavy twine or ribbon into a safety pin or paper clip and thread it through the hem at the top. Tie the two ends of the twine together in a knot. Draw the string to close the bag, thus preventing the marbles from being lost.

Decorate the surface of the bag with patches of fabric, felt pens with permanent ink, or stitchery. If the fabric used for the bag is printed, it may be left as it is.

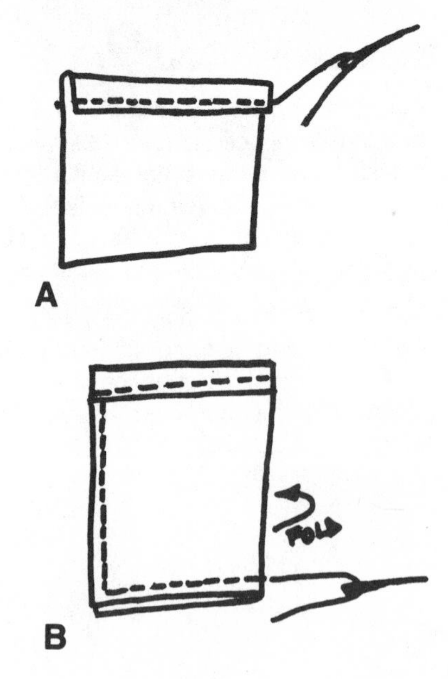

Toss It

Beanbag toss games are fun for young children to play, and beanbags are also fun to make. Simple squares or animal and bug shapes can be filled to make tossable treasures. Even the back of an old worn-out shirt can be put to use in this project.

MATERIALS

fabric scraps
needle and thread
scissors
white glue
beans or dried peas

PROCEDURE

Cut two matching shapes. These can represent objects, such as animals, or they can be purely nonobjective forms. Keep them reasonably small (under 6″) to be really usable. Put the right sides of the fabric together and stitch with small, close stitches (about ¼″ to ½″ from the edge). Leave a one-inch opening on one side for filling. Turn inside out. Fill with dried beans or peas and sew up the opening.

Add any cut fabric details with white glue. Buttons, etc., may be sewn on for additional decoration.

Toss it! A target may be designed for scoring if desired.

Heads Down!

A patchwork pillow can be designed from a variety of fabric scraps cut and then sewn together with a simple stitch. This touch of Americana can be related to a study of early American designers.

MATERIALS

fabric scraps
needle and thread
polyfoam stuffing (optional)

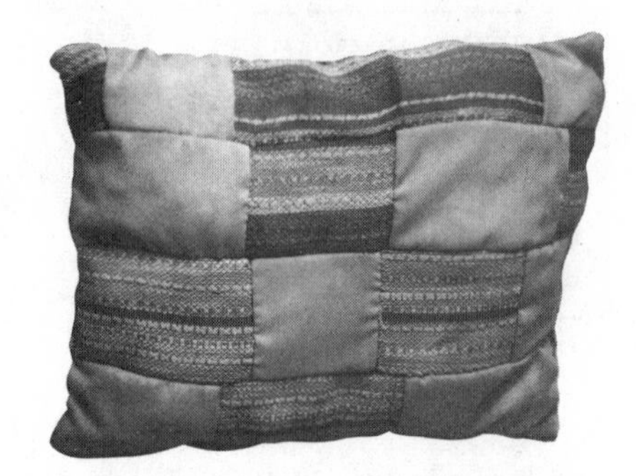

PROCEDURE

Cut fabrics into squares or rectangles, with the sides the same width so the edges will be easier to fit together. Put the right sides of the fabric pieces together and stitch with very small, close stitches about ¼″ from the edge. Stitch enough pieces to make a strip the length of the pillow size (see Illustration A).

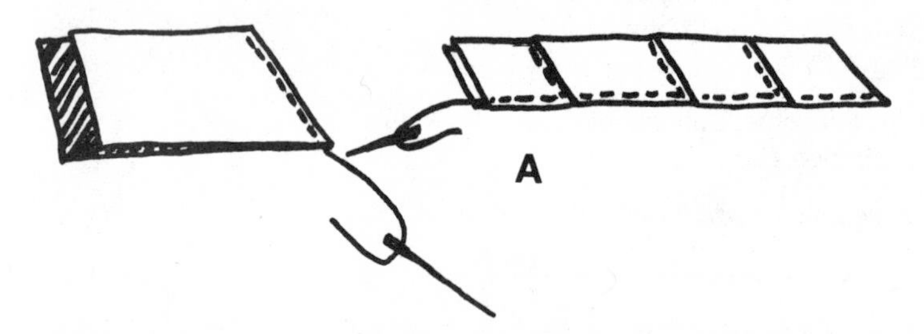

Keep making strips in this manner until you have enough to make a square or rectangular pillow cover of the particular size desired. The back side of the pillow can be made from one piece of fabric or with patches in the same way as the cover.

To put back and front together, put the right sides of fabric together and stitch almost all the way around, leaving a small opening to fill with scraps of fabric or polyfoam stuffing (see Illustration B).

Turn right side out. Fill. Stitch up opening.

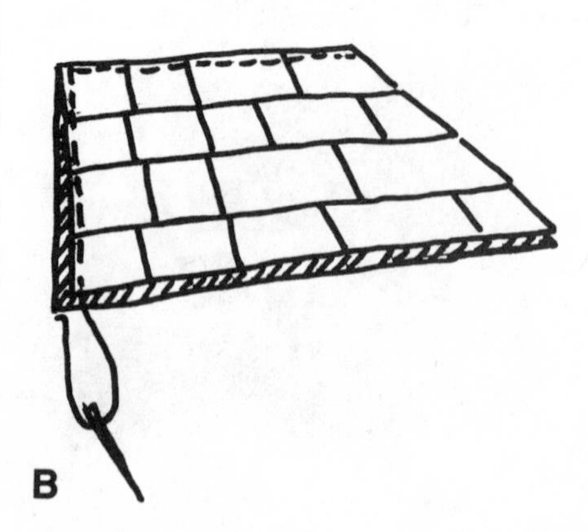

Foot in Your Mouth

Puppetry has long been a craft enjoyed by both young and old. It presents another opportunity for combining the dramatic and the visual arts.

MATERIALS

old sock
needle and thread
white glue
variety of scrap materials (buttons, felt, fur, etc.)

PROCEDURE

Cut a slit in the toe of the sock, which will be the mouth of the puppet (see Illustration A).

Cut an oval to fit the slit in the toe using a scrap of fabric (felt is good for this). Stitch the mouth in place using any kind of stitch. A simple whip stitch is sufficient (see Illustration B).

Add features and characteristics with the variety of scrap materials collected. These may be sewn or glued into place. How about eyes, ears, glasses, or even a hat?

Put the hand into the sock with the thumb in the lower lip and the fingers in the upper lip.

Then, talk your hand off!

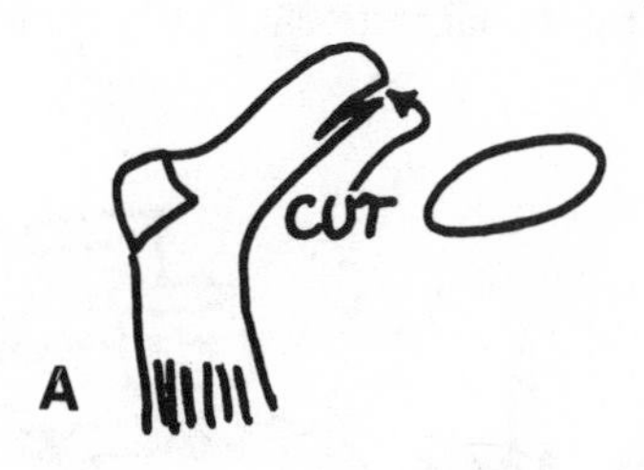

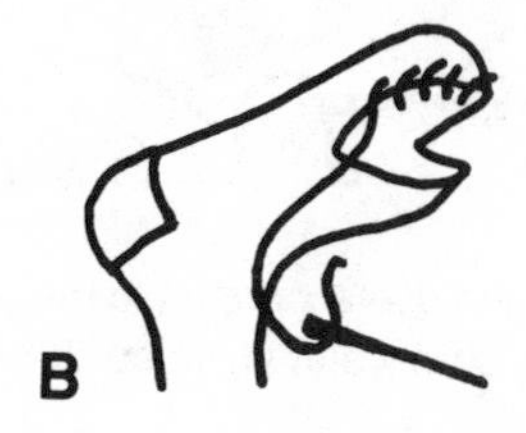

Tie and Dye It

Tie-dyeing is a craft that can be found in Africa, in America, and many places in between. It is an extremely popular craft among young people. Old sheeting or worn and faded clothing can be revitalized with splashes of color. Hang it or wear it, tie-dyeing is a fun experience!

MATERIALS

fabric (sheeting or clothing)
rubber bands or twine
cold-water dye
brushes
newspapers
containers for dye (preferably plastic with lids)

PROCEDURE

Fold or gather the material in bunches and fasten with rubber bands or tie securely with twine. Be sure to tie very tightly because this is what prevents the dye from reaching certain parts of the fabric (see Illustration A).

A

When the *entire* fabric has been tied, the dye application begins. By applying the dye with brushes, you can control color in different areas. The alternate method is to dip the fabric in each dye container (see Illustration B).

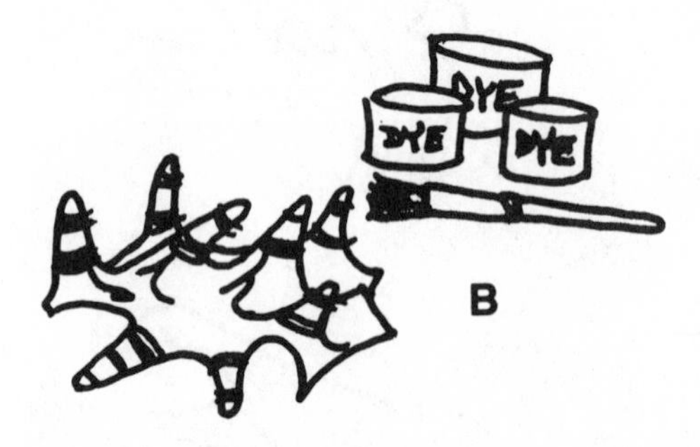

B

Mix the dye according to the directions on the package. Use cold-water dyes so that heating won't be necessary.

Cover the tables and/or floor with many layers of newspaper to avoid staining. Lay the fabric out on the newspapers, and with a *full* brushload of dye begin to saturate the fabric with color. If the weather is right, working outdoors is advisable.

Let the fabric dry before removing the rubber bands or strings. If it's an article of clothing, set the color according to dye instructions. If it's a wall hanging, setting the color is not necessary.

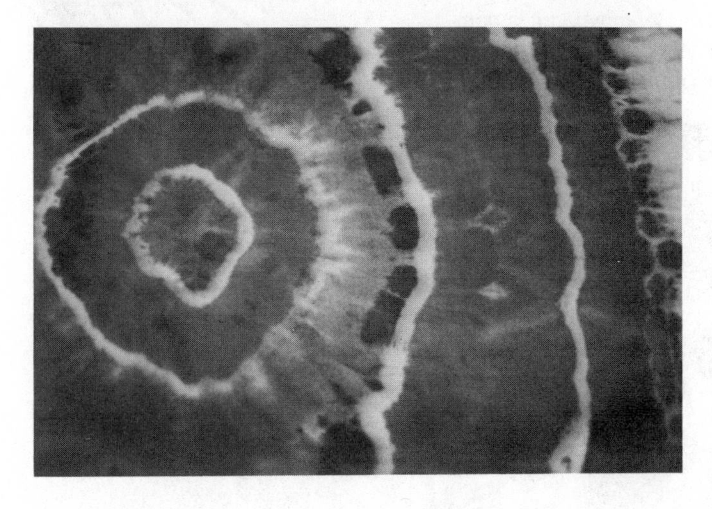

A Panel of Memories

Save those ribbons from Christmas packages for an after-the-holidays project. There are so many things to do before the holidays in the arts and crafts, but what a letdown when it's all over! Here is a project suitable for all ages. To help remember the fun of those holiday activities, weave a panel of memories!

MATERIALS

ribbons and strings from gift packages
twine, yarn, or carpet warp
cardboard box (gift box is good)

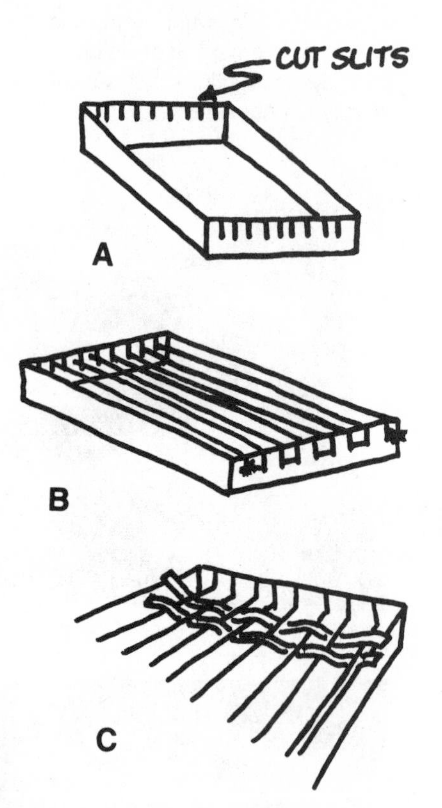

PROCEDURE

Prepare the loom (box) by cutting slits evenly spaced (½″ to 1″ apart) on two ends of it (see Illustration A).

String the warp threads (those running *lengthwise*) by going from one slit to the one across, then to the next one and across again. Continue until all the slits are filled (see Illustration B).

Now the loom is ready for weaving. Cut the ribbons a little longer than the width of the loom. Begin going over and under the warp threads (see Illustration C). Pack each successive row tightly against the previous one until the loom is completely filled.

Slide the panel off the cardboard loom when it is completed. Hang or lay it on a table as a mat. Remember last Christmas with this panel of memories.

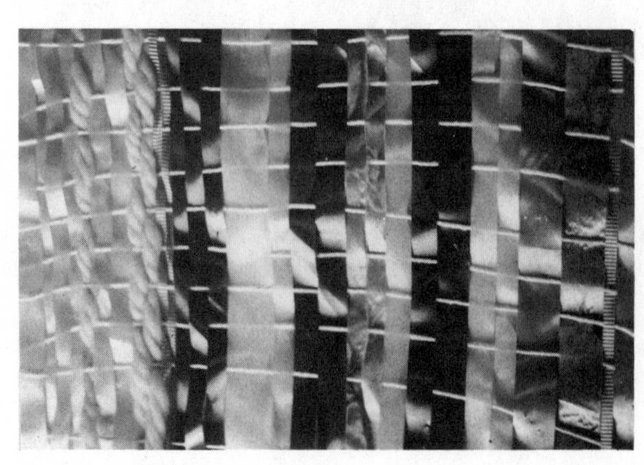

Puffed Ornaments

Stuffed and puffed ornaments add dazzle to a traditional Christmas tree. They go with strung popcorn and cranberries. There's no limit to the colors and shapes that can be used to make these hanging delights.

MATERIALS

scrap fabrics
straight pins
needle and thread
scissors (pinking shears are helpful)
paper
beads, buttons, yarn

PROCEDURE

The shape for the ornament can be cut from paper first and used as a pattern for cutting the fabric. Size should be under 6″ for practical use on a tree. Pin the paper pattern on the fabric, which has been folded with the right sides together (see Illustration A). Cut both layers at one time.

Keeping the right sides of the fabric together, stitch with small, close stitches, leaving a small opening on one end for filling (see Illustration B).

Now, turn it inside out so the right sides of the fabric are exposed. Stuff with small scraps of fabric, cotton, polyfoam, etc. Then, sew up the open end.

Decorate the surface by sewing or gluing on buttons, beads, or yarn. Add a loop of string or yarn at the top for hanging.

Hang it on your tree proudly, knowing that there isn't another one like it anywhere.

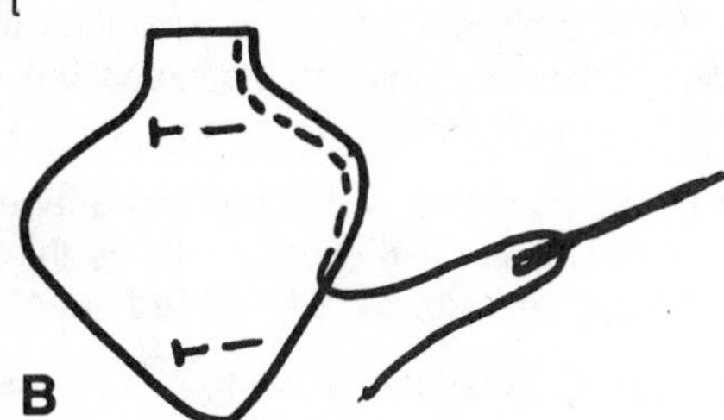

Press It On: Pseudo-Batik

The ancient art of batik, found in many parts of the world, originated in Java. It is a process of developing a design on fabric using wax and dye. Old broken crayons can be used in a kind of imitation batik on worn-out sheets or pillowcases. The resulting banner, picture, or pillow cover has a charm similar to that of the true batik from Indonesia, Africa, or America.

MATERIALS

broken crayons
permanent ink felt markers
old sheets or pillowcases
paraffin
wax melter (muffin pans, juice cans, etc., and electric frypan)
***old* brushes**
India ink
iron
newspapers (lots)

PROCEDURE

Put water in the electric frypan (use low heat). Set the muffin pan or juice cans in the water and melt the crayons in them. The addition of about ½″ cube of paraffin to each color makes a better wax blend than pure crayon.

Draw a design on the cloth with a black felt pen (be sure it is permanent ink). Paint in the areas enclosed by the black lines with colored wax. When completed, crumple the waxed cloth to create a typical crackle effect. This is what makes it look like batik.

Lay the cloth on newspapers and paint it with India ink or dark liquid dye. Rinse off the excess ink by holding under running water.

Iron the fabric between layers of newspaper or paper towels, changing them occasionally as the wax melts out. Do this until all wax is absorbed out of the fabric.

The finished piece can be framed, hung as a banner with sticks, or made into a pillow cover.

Fuzzy Cover-Up

Loose-leaf notebooks seem to have a life of one school year and then wear out. Perhaps with a new look (and feel), your pupils can use their notebooks through another term. Scraps of carpeting or carpet samples glued to the cover can give new life to the guardian of those important school papers.

MATERIALS

carpet scraps
sharp knife (e.g., X-acto)
white glue
felt pen
bristle brush

PROCEDURE

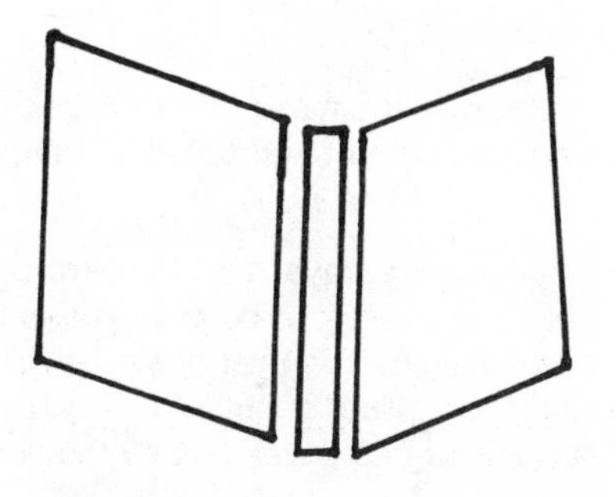

Lay the notebook on the back side of the carpet and draw around one cover at a time using a felt pen.

Cut three pieces—one for each cover and one for the spine (see illustration). Use an X-acto knife or similar knife—*with supervision*.

Apply white glue to one entire cover at a time, using a bristle brush to spread it evenly. Glue the precut pieces of carpet onto the notebook. Allow to dry.

If desired, a pattern of small shapes could be assembled using different colors of carpeting, fitting them together like a mosaic. If this technique is used, cut each piece to make it fit against the next.

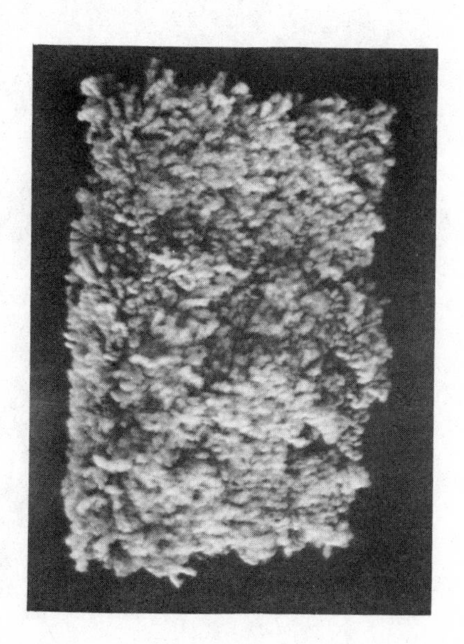

Button Blooms and Heads

Encourage your students to collect some very large buttons. Large buttons could start their imaginations working towards creating flowers or heads in a drawing or painting. Collage, or using pieces of different materials in a painting, goes back to the cubist work of Picasso and Braque in Paris. Today artists still use this technique in varying degrees.

MATERIALS

large buttons
paper
tempera paint or crayons
brushes
white glue

PROCEDURE

Place several buttons on the paper. Try moving them around in different groupings. Then, glue them to the paper with white glue.

Proceed with crayons or tempera paint to develop a picture of flowers or people. The buttons may be the whole flower, the center of the flower, a head of a person, or an insect or butterfly. Anything the shape may suggest can be developed by the creator.

Voodoo Doll Pincushion

Voodoo may not be the real purpose of these little dolls, but they will hold a lot of pins. Give them a human shape or an imaginative one, depending upon the age, interests, and abilities of the children involved.

MATERIALS

scrap fabric (large or small)
scissors
yarn
needle and thread
cotton or other material for stuffing
buttons, etc., for details

PROCEDURE

Cut a shape for the doll out of scrap paper to use as a pattern.

Lay the pattern on a piece of fabric folded with the right sides together. Pin the paper through both thicknesses of fabric. Cut around the pattern about ½" from the paper (see illustration).

With the right sides of the fabric still together, sew almost all the way around staying about ½" from the edge. Leave a small opening at the top or bottom for stuffing.

Turn it right side out. Use the eraser end of a pencil to poke the fabric back in the ends of small arms or legs.

Use colored yarn for stitchery designs and details. Sew on buttons, fabrics, feathers, or fur for enrichment.

Stuff through the opening with cotton, dacron pillow stuffing, or shredded foam. Sew up the opening.

Use as a pin cushion with a character of its own.

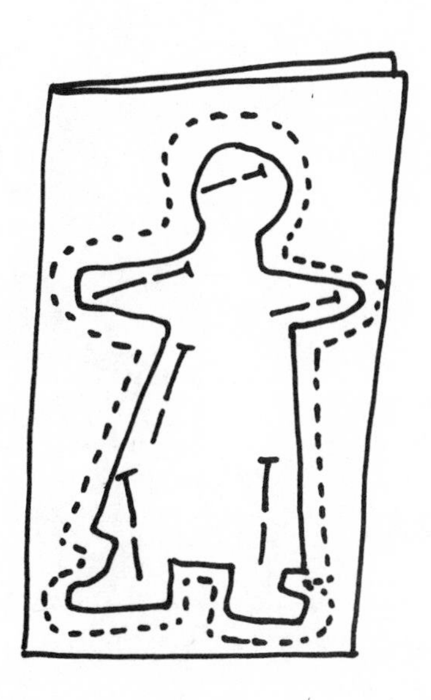

Plastic Lids & Bottles

Pill Bottle Mobile

Free-form shapes appear as plastic pill bottles are melted. The resulting shapes can be hung together as a mobile or as wind chimes. The delicate sound made by these little clear objects moving in the breeze is very pleasant. You should do the melting for younger pupils, but the older ones could do it themselves under close supervision.

MATERIALS

plastic pill bottles
oven (kitchen or portable)
aluminum pan (e.g., frozen pie pan)
string or thread
woodburning tool or nail and pliers

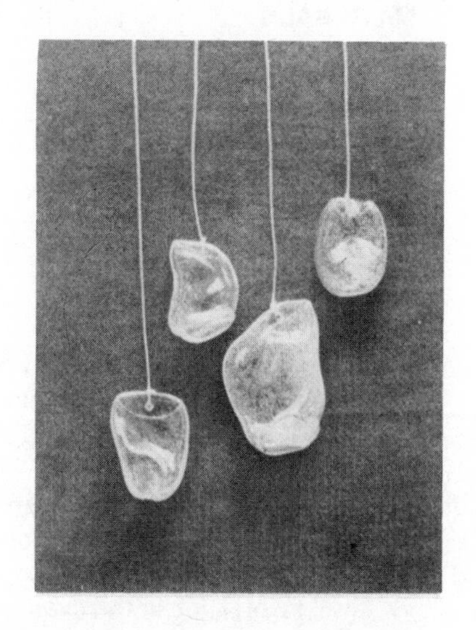

PROCEDURE

A kitchen oven or portable broiler/toaster oven is needed to make these melted bottle shapes.

Preheat the oven to 500°.

Place the plastic pill bottles on their sides in an aluminum pan.

Put them in the oven and watch them melt into new shapes. This may take from 3 to 10 minutes depending upon the plastic. Avoid leaving them in too long, since this causes the plastic to bubble and get cloudy.

Remove and cool. When they have cooled they will come off the pan easily.

Make holes in the top of each piece with the hot point of a woodburning tool, if available. If not, a nail held with pliers can be heated to accomplish the same thing.

Tie strings onto each piece and assemble by attaching them to an unmelted pill bottle with many holes in it, or tie them to a stick. If it is to be hung as a wind chime, be sure that the pieces will touch one another.

Alternative: *Pendant*. One of the melted bottles could be strung on a chain, string, or thong as a pendant. If the top has not melted shut, tiny dried flowers could be inserted for additional color. Leave them sticking out as in a vase, or put them all the way inside so they can be seen through the clear plastic.

Shrink Designs

Clear plastic lids (crisp kind), such as those found on liver or potato salad containers, are a basic material for creating key fobs, necklaces, or wind chimes. Soft plastic coffee can lids won't work for this project. Due to the heat involved, this project is better for older children, but they should use the oven only under supervision.

MATERIALS

clear plastic lids
permanent ink felt markers
oven (kitchen or portable)
woodburning tool or nail
aluminum pie pan

PROCEDURE

Draw a design, letter a phrase or whatever, using permanent ink felt markers in any colors.

Preheat the oven from 350° to 500°.

Place the lids on an aluminum pan and put them into the oven. It takes from a few seconds to several minutes for the lids to melt (the hotter the oven, the less time it will take). If left in the oven too long, the plastic will get milky with bubbles. As soon as it has flattened out and shrunk, remove it from the oven and allow it to cool. Once in a while the plastic will stick to itself when it curls up. If this happens, just start again rather than trying to get it apart. The plastic is very sticky when it is melted.

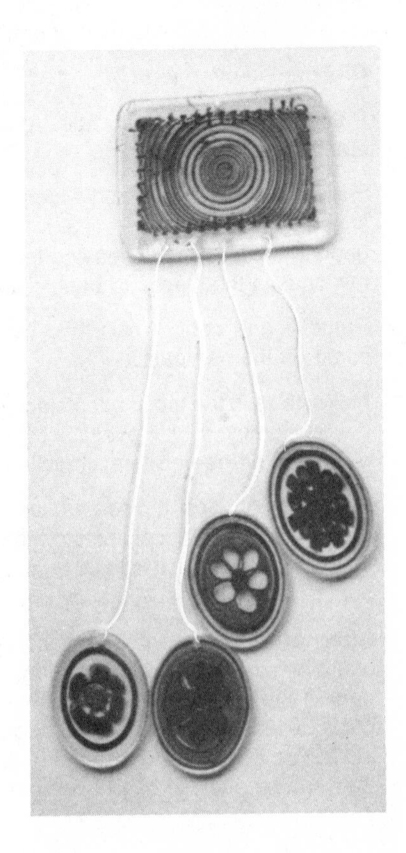

To put a hole in the medallion for hanging, use the heated point of a woodburning tool. If this is not available, use pliers to hold a nail for heating. It works fine.

Place the finished medallion on a key chain or on a string or leather thong for a pendant. Several of these could be hung together from fishing line attached to a piece of driftwood for a wind chime. If they are to be used as a wind chime, be sure that the pieces will touch one another as they move in the breeze.

Fancy Bottle Vases

A well-shaped glass bottle will make a beautiful vase when decorated with colorful tissue paper or colored pictures from magazines. Children of all ages can make these, possibly as Mother's Day remembrances.

MATERIALS

bottle
colored tissue paper or colored magazine pictures
white glue
shellac or gloss (clear) acrylic medium
brushes

PROCEDURE

Choose a bottle with an interesting shape.

Tear or cut shapes from colored tissue paper or colored pictures from magazines. Or, for a still more varied effect, combine both tissue and magazines.

Brush on white glue or gloss acrylic medium to hold the pieces of paper to the bottle. They should be overlapped as they are applied and smoothed out by brushing over them with more glue.

When the bottle is completely covered, allow to dry thoroughly. Then, apply one or two coatings of shellac or gloss acrylic medium.

Bottle People

Plastic bottle people, in varying degrees of complexity, can be created by a child of any age. All that is needed for each one is a styrofoam ball for the head and a plastic bottle with a small neck, such as a detergent bottle, for the body. With these materials and some imagination, your pupils can create bottle people with unique personalities and costumes.

MATERIALS

plastic bottle
styrofoam ball (2″ to 2½″)
facial tissue
white glue (diluted 50% with water)
tempera paint
brushes

PROCEDURE

Take the screw cap off the bottle. Force the styrofoam ball onto the neck of the bottle by pressing down and twisting as though you were screwing on a lid. This helps to hold the head in place.

Tear facial tissue into pieces. Brush on the glue mixture, lay the tissue on the bottle, and brush more glue over it. The tissue will wrinkle and give a rather textured surface. Repeat for 2 or 3 layers. Allow to dry completely.

Paint with different colors of tempera paint to create a suitable costume to fit the character you have in mind. Other materials such as yarn, fur, buttons, or fabrics may be added for interest.

In the photograph you can see the development of the figure from bottle to completed piece. Only paint has been used in this Japanese doll.

Stitch It on the Lid

The soft plastic lids from coffee cans are the prime material for this circular stitchery project. Scrap yarns and thread are used to decorate the lids with colorful geometric designs.

MATERIALS

coffee can lids
yarns and/or thread
needle
scissors

PROCEDURE

Thread the needle with a little more than an arm's length of yarn. Proceed to stitch in and out of the plastic lid, creating a design in one or more colors.

Make a loop on the back for hanging. Hung in groups of two or three, these are very attractive decorations for any wall.

Plastic Lid Mobile

Mobiles are sculptures that move in space, and therefore continually change in design. Everyone enjoys watching mobiles and many businesses make use of them in their promotions. Look for them at the local market.

Have your pupils collect plastic lids from liver, potato salad, and other similar containers. A look at some of Alexander Calder's mobiles will provide good motivation for this project in mobile design.

MATERIALS

clear plastic lids (crisp type)
colored tissue paper
scissors
needle and thread

PROCEDURE

Cut shapes of tissue paper to fit inside the plastic lids. One or several colors may be used and overlapping will make new colors.

Place the pieces in the lid and put another lid of the same size on top of it. By pressing it in all the way around the edge it will snap together. No gluing is necessary.

String several of these lids together using a needle and thread. Hang them from light fixtures or the ceiling and watch them move as the air currents in the room move (see Illustration A).

Alternatives: The lids could be suspended from bent hangers if you want more movement (see Illustration B).

Mobiles also make very attractive and unusual Christmas ornaments, either hung on a tree or by themselves in front of a window.

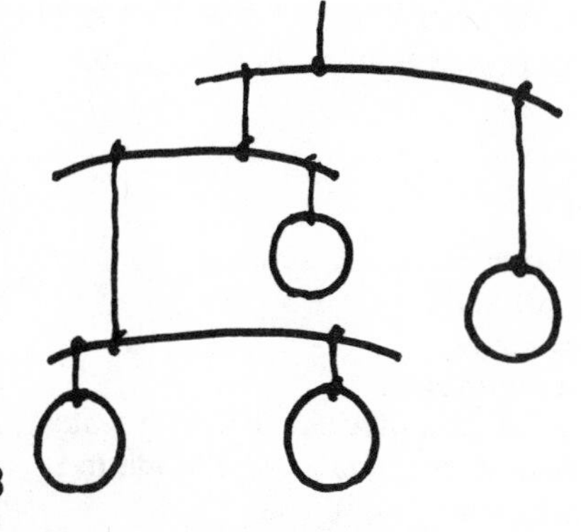

B

Plastic Bottle Bank

You can bank on this one! This will be a very good as well as useful project for teaching your pupils how to save money.

MATERIALS

bleach bottle (or similar)
craft knife
scissors
scrap fabrics
white glue (diluted 50% with water)
shellac or gloss acrylic medium
brushes

PROCEDURE

Cut a slit just large enough for coins using a craft knife. Supervise this part of the activity carefully because of the very sharp blade.

Cut scrap fabrics into small squares. Brush on the mixture of glue and place the fabric squares on the bottle overlapping each one slightly. Saturate the fabric well with additional glue. Proceed in this manner until the bottle is completely covered with fabric squares. If you want to be able to open the bank to empty it, cover the cap separately.

When completely dry, brush on a coating of shellac or acrylic medium. A second coating may be necessary depending upon how absorbent the fabric is. But allow the first coating to dry completely before the second one is applied.

Now start saving!

Yarn-Covered Vase

The Huichol Indian craft of yarn painting can transform an ordinary bottle into a beautiful flower or weed container. These Mexican indians use beeswax as an adhesive, but for our purposes white glue can be used.

MATERIALS

bottle
white glue
newspapers
yarn scraps
brush
scissors

PROCEDURE

Select a bottle with an interesting shape, such as a wine, syrup, or soft drink bottle.

The design may be either a simplified object (flowers, insects, etc.) or purely nonobjective. Limit your color scheme to one light, one medium, and one dark color.

Pour a small amount of white glue in a low container for easy use.

Brush a heavy coating of glue on a small area of the bottle at a time.

Cut the yarn in manageable lengths (about 24″ maximum). Usually, starting at the outside edge of a shape and working towards the center is easier, but not necessary (see illustration).

Be sure all ends are glued down securely. When you run out of yarn, just begin again with a new length at the same point where you left off. If ends are securely glued, the joints won't show. *Hint*: A damp sponge or cloth will be handy to clean glue off your fingers as you work. Sticky fingers will make the yarn difficult to handle. Cotton or wool yarns work more easily than acrylic yarns.

Bottle Baskets

Easter baskets, May Day baskets, or baskets to hold posies can be cut and assembled from discarded plastic soap bottles. Caution: *When students are working with the sharp knife necessary for the project, they should be well supervised.*

MATERIALS

soft plastic bottle (any size—detergent, bleach, etc.)
craft knife (X-acto or similar)
paper puncher
stapler
used gift ribbons or yarns
acrylic or enamel paint (optional)

PROCEDURE

Using a craft knife, *carefully* cut across the front and back of the bottle as shown in Illustration A.

Then, cut up the sides (see Illustration B). The sides will form the handle of the basket.

Staple the two side strips at the top to form the handle. Overlap them slightly for strength.

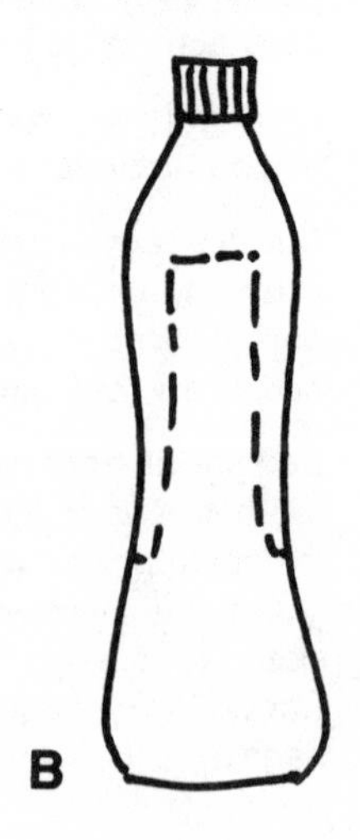

Using a paper puncher make holes approximately ½″ apart around the edges of the basket (see Illustration C).

At this point the basket may be painted with enamel or acrylic paint. This is not really necessary, since most plastic bottles can be found in colors other than white.

Lace used gift ribbons and/or yarn in and out of the holes around the basket. Beads may be threaded onto the yarn here and there for added variety.

Tie a bow on the top of the handle (optional). This will also cover the staples.

The completed baskets can be filled with Easter eggs, candy, flowers, or plants.

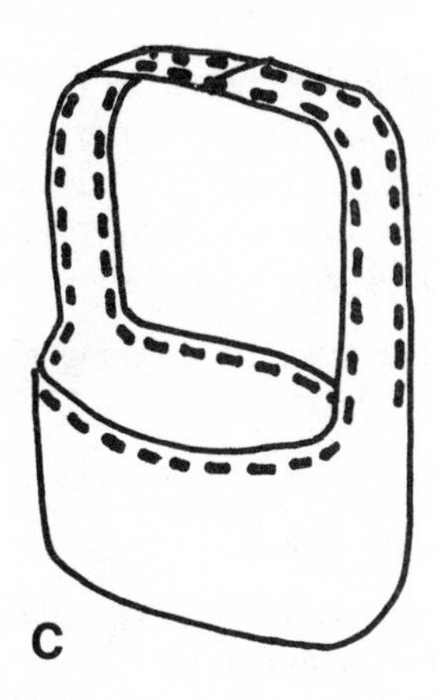

C

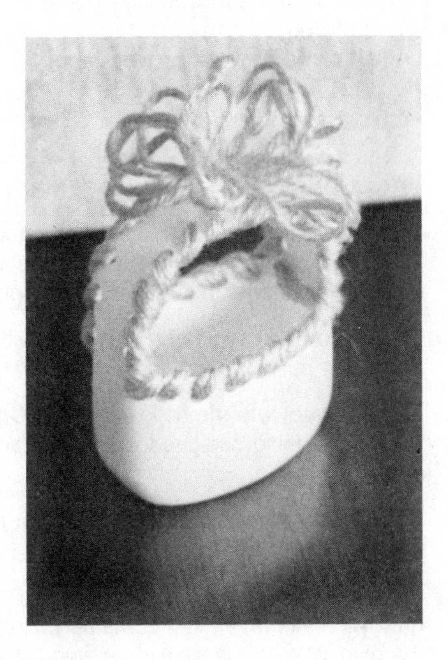

Hanging Floral Display

In today's world of disposable plastic glasses and dishes, we only need to look in the trash can after a party or reception to find many forms to assemble or reassemble into display units or sculptures. Plastic champagne glasses, for example, make beautiful containers for dried flowers or Christmas miniatures. Hang them and enjoy them in a new way.

Because of fumes given off by plastic glue, close supervision of the gluing step is advised.

MATERIALS

disposable plastic glasses (especially champagne)
plastic glue (type used for plastic model building)
small dried flowers or weeds, or Christmas miniatures

PROCEDURE

Wash and dry the glasses. Champagne glasses are the best type for hanging ornaments because they have hollow stems in which to place flowers.

The feet on the stems of the glasses will usually come off by just pulling them.

Place flowers into the hollow stem. If Christmas ornaments are being designed, use small angels, Santas, angel hair, etc.

Then, apply a very thin layer of plastic model glue all the way around the lip of one glass.

Place the other glass upside down on top of the first (this will make an airtight container). Hold it firmly until the glue is set. The glue dries very quickly and securely.

When the glue is dry, tie a ribbon, string, or yarn to the top for hanging individually or in groups. A matching ribbon may also be glued around the middle to cover the seam where the two glasses are glued together.

If you don't want to hang the finished piece, put the foot back on the bottom glass and use it as a stand.

Yarn & Thread

Yarn Weaving on Wood Scrap

Your pupils can make beautifully textured wooden wall plaques decorated with woven yarn ends. Mother, grandmother, or aunt can save pieces of yarn left over from a knitting or stitchery project. Dad can provide wood scraps and nails for the loom. The resulting plaque is not only a great visual experience, but a tactile one as well.

MATERIALS

yarn scraps
scissors
white glue
hammer
nails with heads
wood scraps
pull ring from drink can

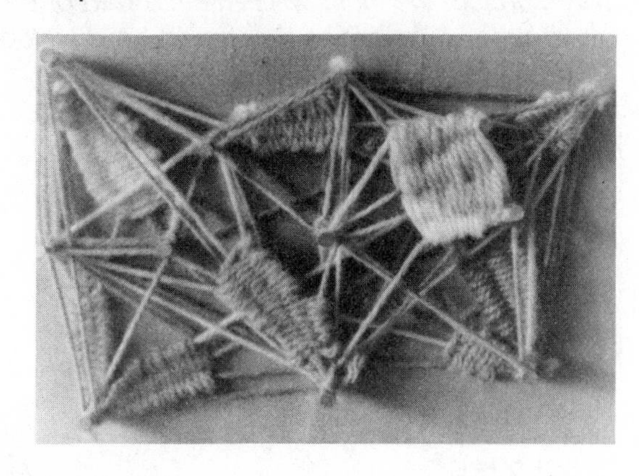

PROCEDURE

Find a scrap of wood. It can be regular or irregular in shape. Use it with the natural finish or paint it if you wish.

Drive several nails at random intervals to give different spacing for the weaving.

Tie a piece of yarn to one nail and stretch it to another. Wrap it around the second nail one complete time. Then, go to another nail. Continue until you have a network of lines covering the board.

Yarn colors may be changed now and then. Tie the ends to the last nail and put a drop of glue on them so they don't stick out.

Weaving: Use an over-and-under weave. Go over one string and under the next, over the next and under the other (see illustration).

You can do several layers of weaving so the woven sections appear to overlap as the string lines do.

Repeat the same colors in more than one place on the design for harmony.

Staple a pull ring from a soft drink can to the back of the wood for a hanger.

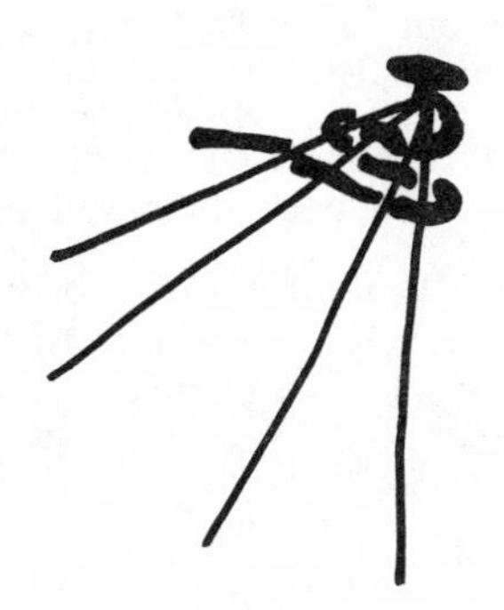

Weaving on Cardboard

The loom is the instrument on which fabrics are woven by interlocking yarns in and out. Fabrics range from very regular, smooth pieces used for wearing apparel to irregular and textured pieces for decoration.

In this project the loom will remain a part of the completed panel. Your pupils will see lots of other possibilities that could be developed as they work.

MATERIALS

cardboard or chipboard
scissors
ruler
string or carpet warp
yarn scraps
ribbon or fabric scraps

PROCEDURE

Cut a piece of cardboard or chipboard into a rectangular or square shape. Size depends upon the designer's ambition, but it's easier to work with one 6″ x 9″ or larger.

Using a ruler, measure ½″ marks along two opposite ends. Make ½″ slits on these marks (see Illustration A).

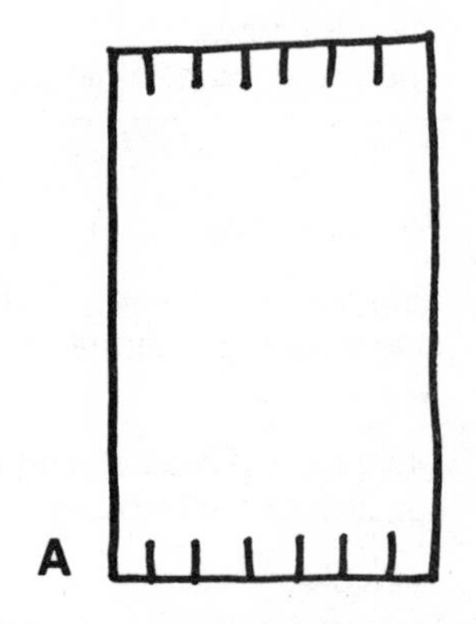
A

The loom is now ready for the warp (the threads which run *lengthwise*).

Tie a knot in the end of a long piece of string or carpet warp. Slip the string into the first slit with the knot on the back side. Then, stretch it to the slit directly opposite. Move to the *next* slit, and bring the string up and back across to the slit directly opposite. Keep the strings tight, but not so tight that the cardboard bends (see Illustration B).

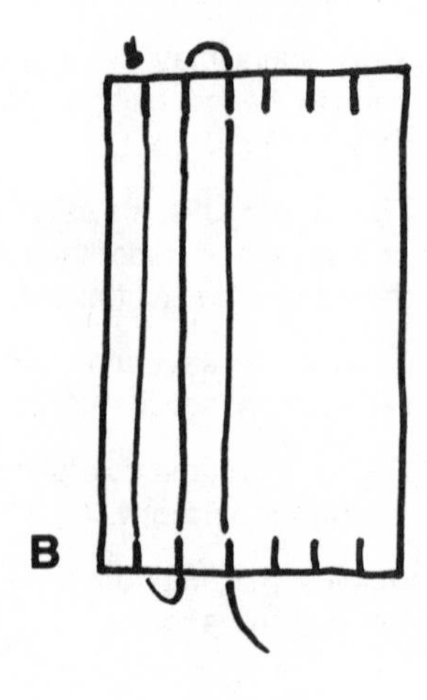
B

Continue back and forth until all the slits are filled. Tie a knot on the back at the final slit.

Now, the warp is ready for weaving. Use scraps of yarn, ribbons, or narrow strips of fabric. These will form the woof or the weft (threads running *crosswise* on the warp).

Weaving. Take the woof over the first warp thread, under the next, over the third (see Illustration C). Continue the over-and-under process to the last warp thread. The woof thread can be cut at the edge of the cardboard, or you can continue with the same piece.

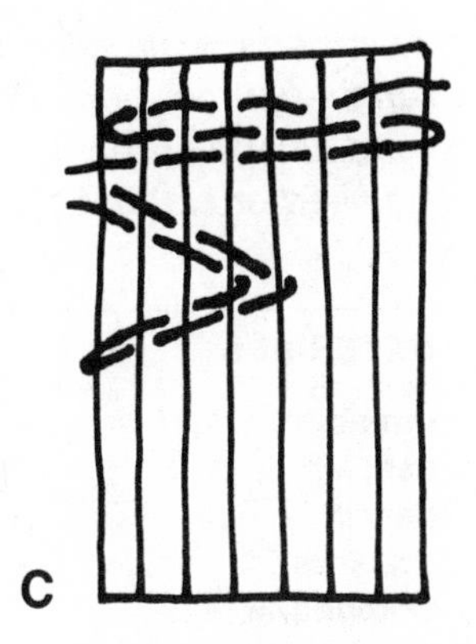

Reverse the direction, going over and under opposite threads the second time across. This will lock them into place. Avoid pulling the woof too tightly or the sides of the loom (cardboard) will pull in.

When you reach the other side again, push the woof threads together.

To create an irregular design, you can reverse direction partway across, or go down at an angle. Experiment to see how you can vary your design. Intermix yarns and ribbons or fabric strips.

When the weaving is finished, the panel can be hung as a wall decoration by putting a loop of yarn at the top. It could also be slipped off the cardboard for displaying, but the cardboard backing gives it extra support.

Stitches on Paper

Sewing on construction paper makes a delicate, sweeping design. Curves develop, even though only straight lines are used in their construction. Partially used spools of thread can be collected over a period of time prior to the big day of stitching. Color can be limited or elaborate, depending upon the supply of thread.

MATERIALS

thread
needles
pencil
scissors
cellophane tape
construction paper or similar (12″ x 18″ approximately)

PROCEDURE

Fold the piece of paper in half (either direction) so you'll be working on a double thickness of paper. Try to use a color that contrasts with the color of the thread available.

On the back side of the folded paper, draw two or three smooth curves (see illustration).

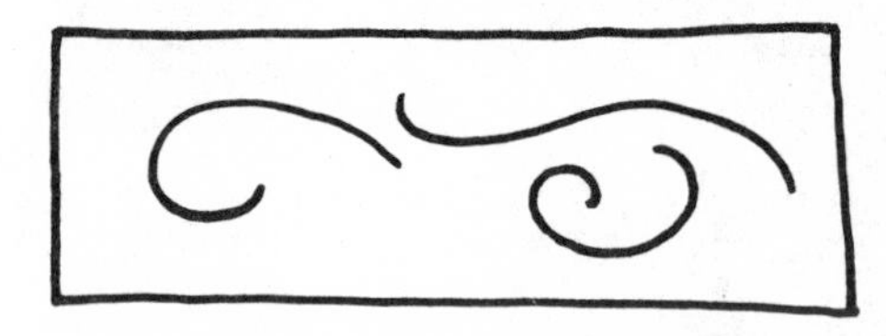

Using a needle, punch holes along the lines about ½″ apart. Be sure to hold the paper tightly so the holes are in the same position on both halves of the paper.

Thread a needle with about two arm's lengths of thread Double and tie the ends in a knot.

Come up through a hole from the *bottom* side. Tape the end of the thread to the back.

Go down into a hole across the front, then up through the *next* hole from the back side.

Build your design in any direction. Threads may cross over one another and the holes may be used more than once.

Each time you end a thread or start a new one, tape it on the back side of the paper.

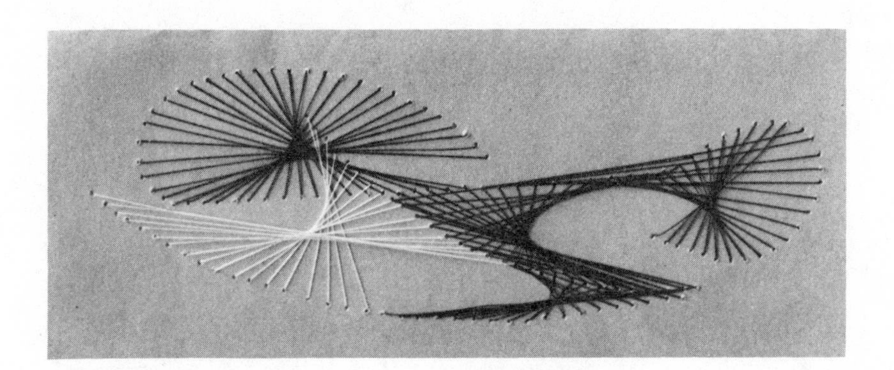

Ojo De Dios (God's Eye)

The God's Eye is a prayer in visual form to protect children and to guarantee long life to the person making it. It is also believed to keep watch over the household. This is a very decorative as well as religious object found in Mexico and Brazil.

MATERIALS

yarn assortment
sticks or magazines
white glue

PROCEDURE

The crossbars for the base of the God's Eye can be made from sticks, twigs, dowels, or paper. Two pages of a magazine rolled together form a very sturdy, inexpensive stick.

To roll a stick, put two pages of a magazine together and roll diagonally on a pencil from one corner. When you have started the roll, remove the pencil. Hold the paper very tightly as you roll to keep the stick small and hard (see Illustration A). Glue the end with white glue. You'll need two of these sticks.

Cross the sticks and tie them together in the middle. Tie from both directions (see Illustration B).

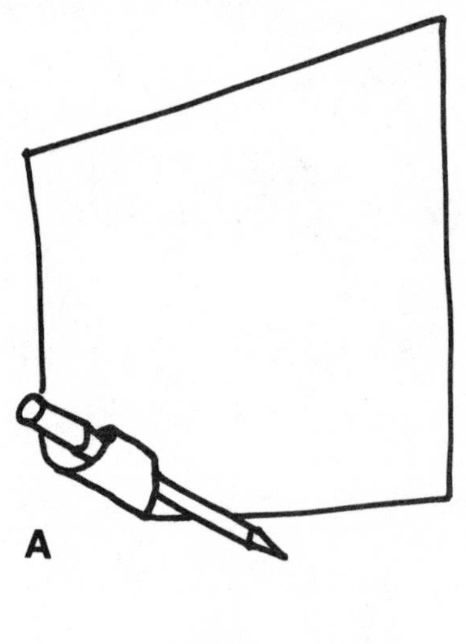

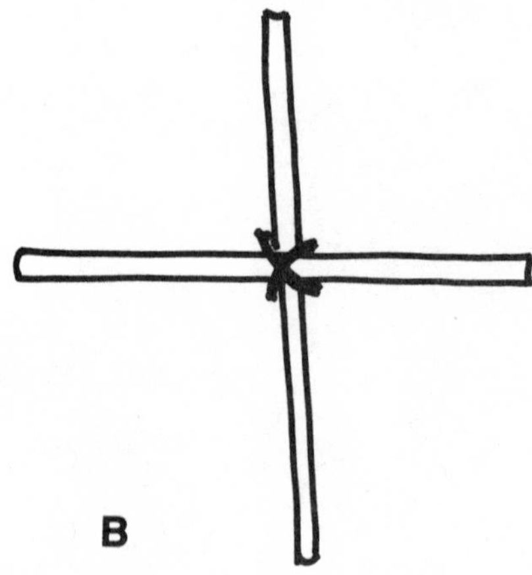

Begin wrapping with yarn at the center (see Illustration C).

Go around stick A clockwise, then to B. Go around stick B clockwise, then to C. Go around stick C clockwise, then to D. Go around D, then back to A. Continue the process. Pull the yarns snugly, but not too tight or the sticks will break and the yarns will sag.

When you wish to change color, glue the end of the yarn to the stick with white glue. Glue on a new color at the same point and continue wrapping.

Tassels may be added, if desired.

Hang singly or in groups of various sizes and lengths.

These are very pretty as Christmas tree ornaments.

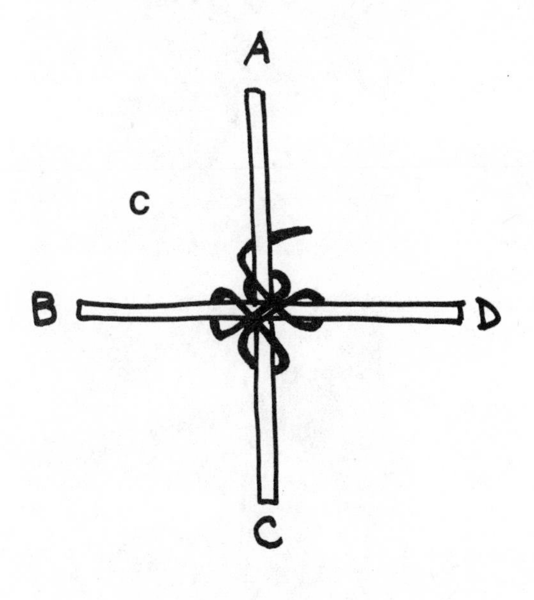

Yarn Painting on Cardboard

Again, we borrow the technique of Huichol Indians of Mexico as the class "paints" pictures with colored yarn scraps. White glue is substituted for the beeswax that the Indians use.

MATERIALS
cardboard or heavy paper **white glue** **pencil** **yarn scraps (preferably cotton or wool)**

PROCEDURE

Draw a design lightly in pencil on a piece of cardboard or very heavy paper.

The yarn may be used to cover the paper completely or some of it may be left blank. Both techniques are shown in the examples.

Cover a small area of the paper with white glue. Lay the yarn on the wet glue creating a raised line. If you are covering the paper completely, begin at the outside edge of the shape and wind the yarn into the center. Press it into the glue as you work to be sure that it is well secured. When you run out of yarn, just begin where you left off with a new piece. A pencil will help you guide the yarn around.

If you are not completely covering the paper with yarn, apply lines of glue with a squeeze bottle.

In either case, apply glue to small areas at a time so it won't dry before it is covered with yarn.

Alternate: Cut small shapes for jewelry or Christmas ornaments. Proceed to cover them in the same way as the pictures.

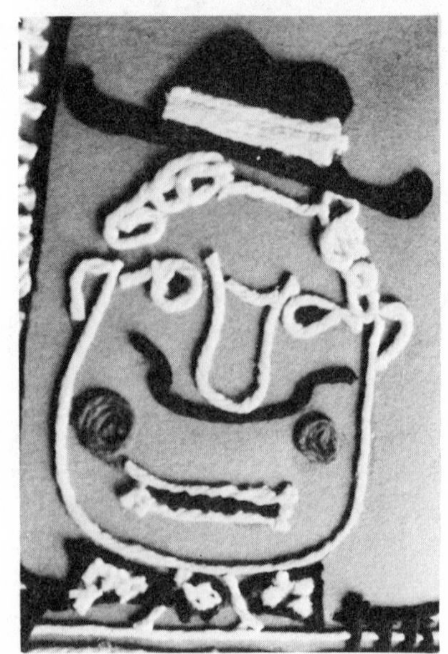

Double Your Pleasure

You can teach your class about symmetry with nothing more than a small length of yarn and a piece of paper. What you have on one side of a sheet of paper will be what you will get as a mirror image on the other side. Twice as much for your effort!

MATERIALS

yarn scraps or string
paper about 12″ x 18″
tempera paint
brush
newspapers
pie pan

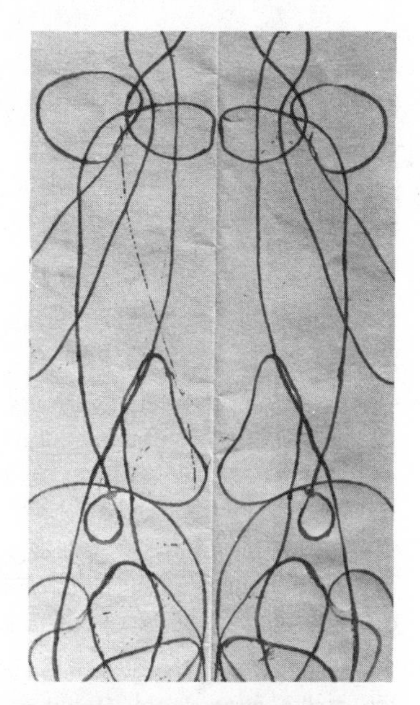

PROCEDURE

Fold a piece of paper lengthwise. Open and lay it flat on newspaper.

Pour a small amount of thin tempera paint in a pan.

Hold the yarn (about 24″ to 36″) by an end and dip the rest into the paint to saturate it. Hold the yarn against the side of the pan with the brush and pull it out. This will remove some of the excess paint from the yarn.

Lay the painted yarn on one side of the fold, making a line design as though drawing with a pencil. Carefully fold the paper on the same fold and rub the full length of it with the palm of your hand.

Open the paper carefully and lift the yarn off. You will have a symmetrically balanced design.

Another color could be added by repeating the same procedure again.

Alternative. Instead of printing a simple line design, you might try the pulled design shown in the second example.

Follow the same procedure, except pull the string out before opening the paper.

This works best if two children work together. Have one child hold the paper folded flat while the second pulls the string straight out from the end. Then, open the paper to find a symmetrical, smeared design. More color and solid area are created by this second technique.

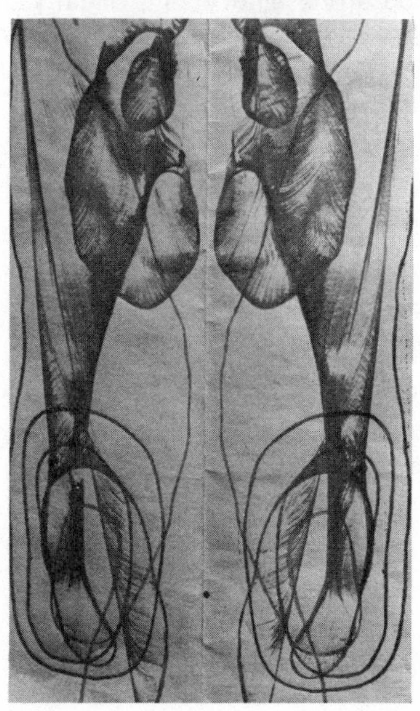

Weave a Belt of Straws

Used plastic drinking straws make a good loom for weaving a belt or sash. Paper straws will do, but the plastic ones are stronger and can be washed more easily. Whether one color or many are used, the result will be a belt, sash, or headband which can be worn proudly.

MATERIALS
yarn
straws
scissors
white glue

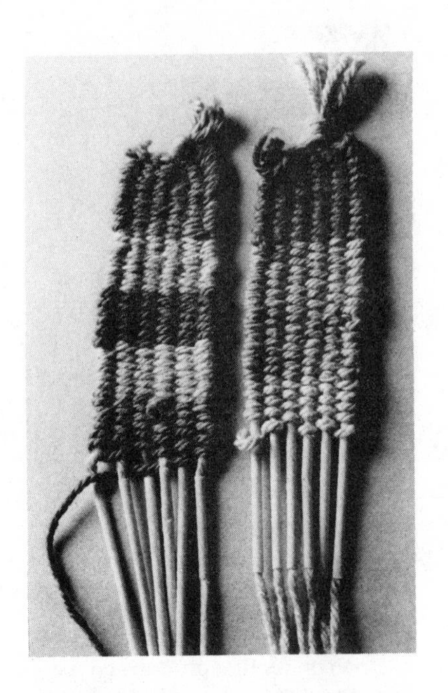

PROCEDURE

Use four to six strands of yarn, depending on the width desired. The length of these strands will determine the length of the completed belt (use about two times your waist size). Tie the strands together with one big knot.

Thread each strand through a drinking straw. A wire may be helpful for pulling it through, or you might suck it through. Push the straws all the way up to the knot.

Tie the first yarn to be woven to the outside string at the top of the straw loom. Begin weaving over the first straw and under the second, over the next and reverse back again when you reach the last straw (see illustration).

When you want to change colors or add more yarn, use white glue to fasten the pieces together. This is neater than knotting.

As the weaving progresses, slide the woven section off to expose more of the loom straw. However, don't slide it all off until you are finished.

At the end, slide off the straws and tie another large knot similar to the one on the other end, forming a tassel.

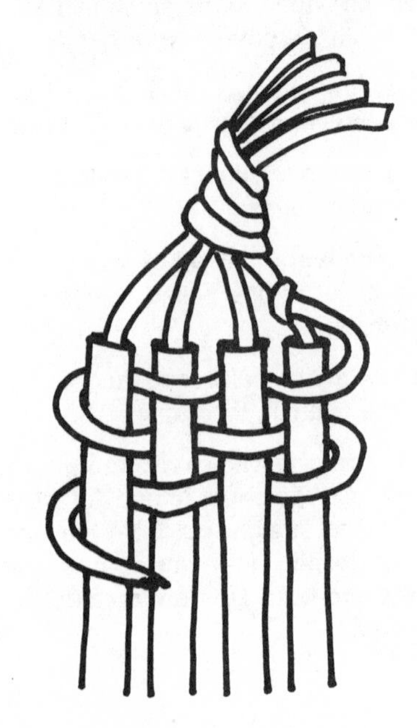

Yarn Ornaments

Christmas ornaments come in all sizes, shapes, and colors, and these lacey beauties are a good addition to any collection. Hang them on a tree or from a light fixture on the ceiling so they can move in graceful spins.

MATERIALS

yarn scraps
waxed paper
potpie pan
white glue (diluted with 50% water)
thread
pencil or popsicle stick
newspapers

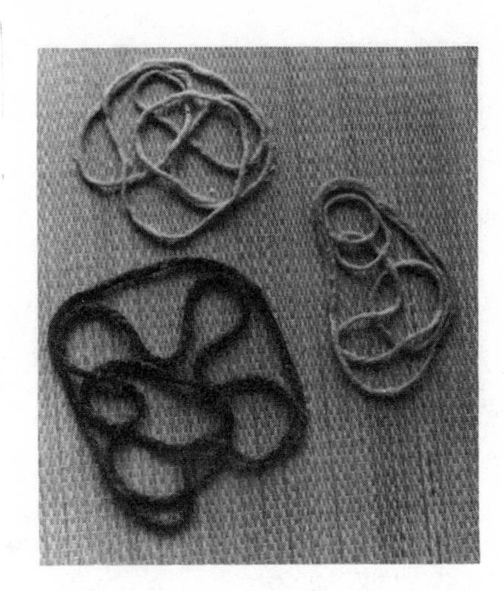

PROCEDURE

Pour diluted glue into an aluminum pie pan. Cut the yarn into 18" to 24" lengths. Put waxed paper over newspapers.

Place the yarn into the glue mixture and saturate very well. Pull it out of the pan holding it between your thumb and finger, but don't squeeze it. Just get the excess glue off the surface of the yarn.

Now, make a line design on the waxed paper. Use the yarn as you would a drawing tool. Overlap the lines for strength. Use a pencil or popsicle stick to slide the yarn around on the waxed paper for shape. The more strings that touch, the stronger the ornament will be.

Allow it to dry overnight without moving.

Next day, peel off the waxed paper and hang with a thread, individually or in groups. Hanging from a tree or near a window, they move gracefully as the air in the room moves.

Spool Knitting

Older children will enjoy wearing a yarn cap or necklace they have made with a tool as simple as a spool. If a cap seems too ambitious, perhaps they might start by weaving a mat using the same basic process.

MATERIALS

thread spool
headless or small head nails (1″ long)
yarn
crochet hook
thread
needle
scissors

PROCEDURE

Drive four nails into one end of the spool an equal distance apart. Leave them sticking up about ½″ to ¾″ (see Illustration A).

Thread the yarn down through the hole in the spool. Hold it at the bottom of the spool. Now, wrap the yarn around the nails as shown in Illustration B. Go around nail A counterclockwise, and over to nail C. Go around nail C clockwise, and over to nail B. Go around nail B clockwise, and over to nail D. Finally, go around nail D counterclockwise.

Now you're ready to knit. Take the yarn to nail A. Hold it against the nail above the strand on the nail. Use the crochet hook to lift strand A over strand B and over the nail (see Illustration C).

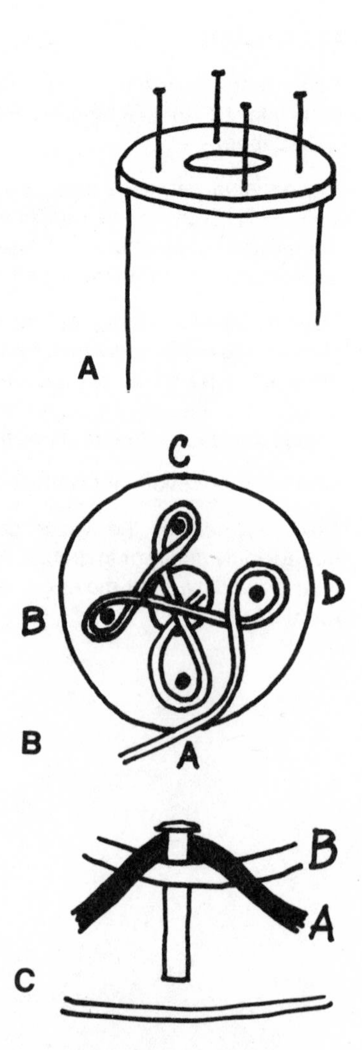

Move to nail B and repeat the same procedure. Continue to nail C and nail D, then go back to nail A. Pull down on the strand sticking out at the bottom of the spool.

Keep rotating until the knitted rope coming out the bottom is as long as needed for the project. You'll need 8 to 10 feet all together, but it doesn't have to be in one piece.

If you are making a necklace, a shorter length is needed.

At the end of the strand, slip the yarn loops off the nails and run the yarn through the holes as they come off the nails.

To make a cap, stitch the rope together in a spiral. Use the same yarn or thread and stitch along the edges (see Illustration D). To make a mat, be sure to keep the spiral loose so it will lay flat.

Add a tassel or a ball to the top to finish the cap.

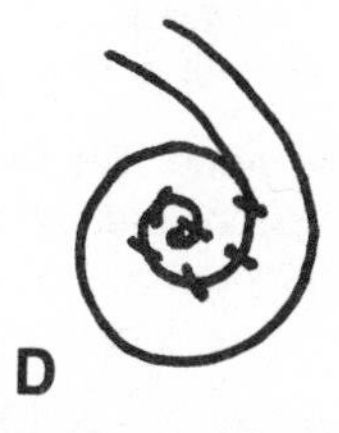

D

String Head

Your students can wind inflated balloons with string to make hanging or standing comical heads. You might have a crowd of these heads hanging from the light fixtures at different levels. They spin and "talk" to each other as they move.

MATERIALS

string or carpet warp
white glue (diluted 50% with water)
construction paper
newspapers
brushes
scissors
balloon

PROCEDURE

Blow up a round or oval balloon. Wrap the balloon with lightweight string or carpet warp going in many directions to create a network of lines. Wrap it gently but tightly, with many overlapping lines (see illustration).

Make a cylinder by rolling a piece of construction paper. This will act as a base and hold the wrapped balloon while it is drying.

Brush the string with diluted white glue, saturating all of the strings completely. Set it on the paper cylinder to dry overnight. Place newspapers under the cylinder in case the glue drips.

The next day, gently push the balloon away from the strings—just enough to loosen the glue. Then, pop the balloon with a pin. Pull the balloon out between the openings in the strings.

Now, cut shapes of paper for features on the face (e.g., eyes, mouth, ears). Add any special decorations, such as hats, glasses, or earrings. The cylinder will be the neck for the head, so you might add a collar or tie to it.

If this is going to be a hanging head, remove the cylinder and add a string at the top. Watch it spin and turn in the air.

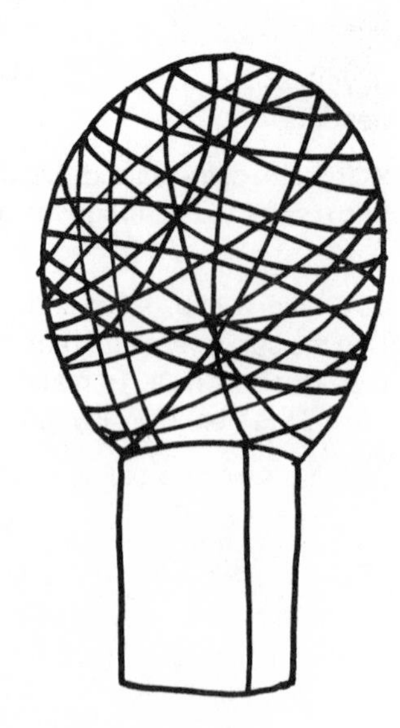

Foam & Foil

Op-Pop Art of Styrofoam

The optical illusions of Op Art and the use of everyday materials in Pop Art combine to make an exciting project. Children find drawing exciting when they use felt markers to create three-dimensional patterns on styrofoam, patterns to be viewed from all sides. The search for interesting styrofoam packing shapes is an important part of the experience. The more highs and lows, or the more "positives" and "negatives" the piece has, the more surfaces there will be for designing.

MATERIALS

styrofoam packing from small appliances (TVs, etc.)
permanent ink felt marking pens (one color)

PROCEDURE

Have your class look at some Op Art to stimulate an awareness of pattern and an interest in creating a similar type of expression. The use of felt markers of only one color will direct the student's attention to pattern and line rather than color. A fine-point pen is better for this project than one having a broad point.

Begin designing at any point, and keep working all around and in and out of the surfaces.

The completed project can be suspended from the ceiling so it will move, or it could be set on a table as a sculpture. Several pieces might be put together to form a large sculpture and then restacked like building blocks to change the design over and over again. How about a whole wall decorated with pieces fitted together to create a high-low relief sculpture?

Adornments

This jewelry may not be found at Tiffany's or Saks, but it will be precious to its designers. Boys and girls can make these adornments for themselves or as beautiful gifts. Styrofoam packing comes in many shapes and sizes from discs, to rings, to cubes. Find them and design unique pieces of jewelry with them.

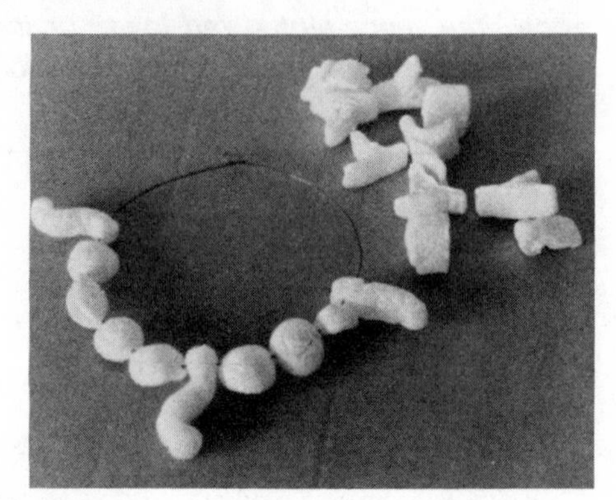

MATERIALS

small pieces of styrofoam packing
wire (if a neck-ring is being made)
string, thread, or fishing line (if a necklace)
needle
permanent ink felt pens (optional)

PROCEDURE

If the styrofoam beads are to be decorated, use permanent ink felt pens or any other permanent coloring material. Avoid watercolors or tempera paints, since they are not permanent and will come off on the clothing when worn.

Necklaces can be made by stringing the pieces on heavy string, thread, or monofilament nylon fishing line with a needle. Long ropes to go over the head or chokers to be tied in the back may be designed.

Neck-rings using a wire form can be strung with the foam shapes as well. Begin with a length of wire long enough to fit the neck. Bend a closed loop on one end, then string on the beads in any pattern desired.

When completed, make an open loop on the remaining end of the wire ring. This will lock into the closed loop at the back. Needle-nose pliers will be helpful when bending these loops (see illustration).

Mondrian in Three-D

Find some examples in art books of Mondrian's geometric paintings in primary colors. These examples will motivate the children to develop their own designs for paintings in three dimensions. The layout can be related to geometry, creating an awareness of squares and rectangles, or you might expand the idea to include any other shapes with angles.

MATERIALS

flat styrofoam boxes
masking tape of various sizes
tempera paint, or permanent ink felt markers of various colors

PROCEDURE

Apply masking tape to the styrofoam box as though you were drawing straight lines. Work for large and small related shapes on top as well as on the sides, carrying the design continuously over the entire surface.

When you are satisfied that the design is complete, begin applying color to the exposed styrofoam surface. Color may be limited to the primaries or to other types of color schemes, e.g., monochromatic or analogous. When the paint is thoroughly dry, carefully peel off the tape.

The completed work may be hung, set, stacked, or mounted on the wall.

Sponge Painting

Introduce your students to sponge painting. Old sponges often have enough life left to serve as painting tools. Natural, cellulose, and foam sponges all have a different character when they are coated with paint and pressed onto a surface. Your class will be fascinated by the textures and patterns created.

MATERIALS

sponges
tempera paint
TV dinner trays
paper (colored construction paper is good)
coffee can for water
newspapers

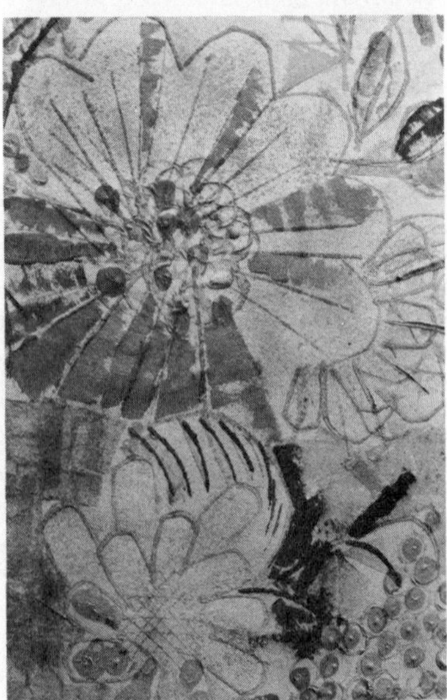

PROCEDURE

Cut the sponges into small pieces about 2″ x 2″ x 2″ or even smaller. Pour small amounts of liquid tempera paint into a TV dinner tray or a similarly divided pan.

Dip the sponge into water and squeeze it almost dry. This will make the sponge more absorbent so it will hold paint.

Dip the sponge into the desired color, just enough to cover the surface. Then, press it onto a piece of newspaper to get some of the paint off. Stamp it onto the paper surface to create your painting. Avoid wiping the sponge on the paper, which will destroy the sponge texture. Use it more as a printing tool than a brush.

Each time a color is changed, use a fresh sponge or rinse the used one well in water.

Other tools may be used to add sharp lines to the design, such as the edge of a scrap of cardboard. Simply dip the cardboard into the paint and press it on the paper. Cardboard bends very easily, so curves can be printed as well as straight lines.

Foam "Soft Sculpture"

In recent years, sculptors have been experimenting with materials other than the traditional clay, tone, and wood. The creation of "soft sculpture" using fabric, foam, macramé, etc., makes an interesting project for pupils of most ages and abilities.

Foam rubber scraps from upholstery or carpeting can be used for this enjoyable three-dimensional experience.

MATERIALS

foam rubber scraps
scissors
twine or carpet warp

PROCEDURE

Twist, cut, and wrap pieces of pliable foam rubber to form animals or purely nonobjective designs in space.

Wrap and tie the foam securely with any kind of strong string. The more wrapping that is done, the stronger the piece will be.

Foam Printers

Nearly every sort of material your class can find will create a printed design when painted and pressed against paper or cloth. In most cases, it is difficult to change the surface of the object for printing; but foam meat trays or the tops of foam egg cartons can very easily be cut and drawn on to create a printing surface.

An edition of pictures, greeting cards, or wrapping paper can be produced by pupils at all grade levels.

MATERIALS

foam tray or egg carton
tempera paint
brushes
newspapers
scissors
pen or pencil
paper for printing

PROCEDURE

Cut a piece out of the tray or egg carton top so you have a flat piece of foam. Some meat trays have a design or number embossed on the bottom. Try to avoid using this because it will print.

Use a pencil or ball point pen to draw a design on the foam. Only slight pressure is needed to make a depressed line in the foam.

When you are making a printing block, always remember that it will print the opposite way. If words or numbers are used, they have to be drawn backwards so they will print in the right direction.

Cover the whole surface of the completed design with tempera paint. The foam sometimes resists the paint. If it does, apply the paint and then wipe it off. Reapply more paint and carefully place the foam upside down on the paper to be printed. The paint will make the foam stick to the paper, so you can carefully turn it over and rub gently all over the back of the paper. This will give a clearer print than just pressing it on.

Now, peel off the foam and repeat the process.

Meat Tray Stitchery

Young children sometimes have trouble doing stitchery on loose fabric. With a foam meat tray as the base, working with yarn can be easy and fun. No elaborate stitches are needed. Just pushing the yarn in and out of the foam will create colorful designs and help to develop coordination.

MATERIALS

large-eye needle with a blunt point
scissors
colored yarn scraps
foam meat tray

PROCEDURE

Thread a needle with yarn—about two arm lengths. Too much yarn will get tangled up. Tie a knot in one end of the yarn.

Begin the design by coming up through the bottom of the tray so the knot will be on the back side. Each time a new yarn is used or a change of color is made, begin in the same way. *Caution:* Don't put the needle holes too close together or the foam will break.

When the design is completed, mount the plaque on colored paper or put a loop of yarn on the back for hanging.

Cup Heads

Styrofoam coffee cups provide a good base for a series of heads. Felt pens and imagination can transform these inanimate objects into many kinds of characters.

MATERIALS

used foam cups that have been washed
permanent ink felt markers (in one or more colors)

PROCEDURE

Turn the washed cup upside down and think, "What kind of person could this be?" As your students look at the blank white cups, their imaginations will help them add eyes, nose, mouth, hair, ears, eyeglasses, and perhaps even a wart!

One color or several colors may be added, but be sure that permanent ink felt markers are used. Water-base ones will smear too easily.

Display the finished pieces in a row or perhaps hang them from a string.

Cup Towers

Foam coffee cups cut and stacked form exciting open towers. Younger children can stack and glue the cups without cutting them. Even older children, when using the sharp tools necessary for this project should be very well supervised and work only in small groups. The towering results will be well worth all the precautions needed.

MATERIALS

used styrofoam coffee cups that have been washed
X-acto knife (or similar craft knife)
tacky glue for foam and plastic

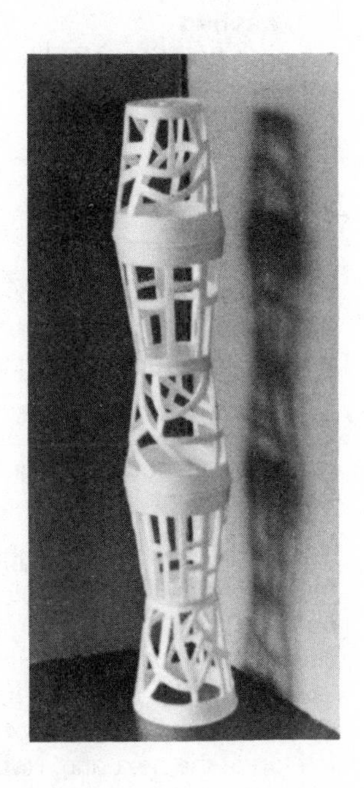

PROCEDURE

Begin cutting openings into the sides of the cup, being very careful to keep fingers away from the sharp blade. It is not necessary to draw a design first, but you may draw one if you wish. Use a very light pencil line so it won't show on the completed piece.

Cut several cups with a design all the way around to create an open filigree effect.

Glue the cups together—bottom to bottom and top rim to top rim. Use the tacky glue made for foam and plastic for best results.

Stack them as high as you want. A firm base may be needed if the tower gets too tall. A block of wood will serve as a good foundation.

Ring Mobile

Styrofoam cups can be cut very easily into rings with variations in general shape. When these are hung with strings, they make an interesting kind of mobile or moving sculpture. Because of the sharp tools necessary, it is better to limit the project to older children.

MATERIALS

styrofoam coffee cups that have been washed
X-acto knife (or similar craft knife)
thread
needle

PROCEDURE

Cut rings from the cups by cutting all the way around the cup in a set pattern. For example, if you begin with a wavy line, continue all the way around with a wavy line. If you begin with a zigzag, continue with the same zigzag all the way around. When you have a number of rings cut out, you can begin to string them for hanging as a mobile.

Thread the needle with lightweight sewing thread. Put the needle down through the top of the ring, and then back up to go down through the same hole again (see illustration). This will form a loop through the ring so it won't slide down on the string. Then put the needle down through the bottom of the ring and continue on to the top of the next ring. Repeat until a line of rings are connected.

Hang several lines of rings together or simply put them next to one another from the ceiling or a light fixture. Light air currents in the room will cause them to move and spin gently.

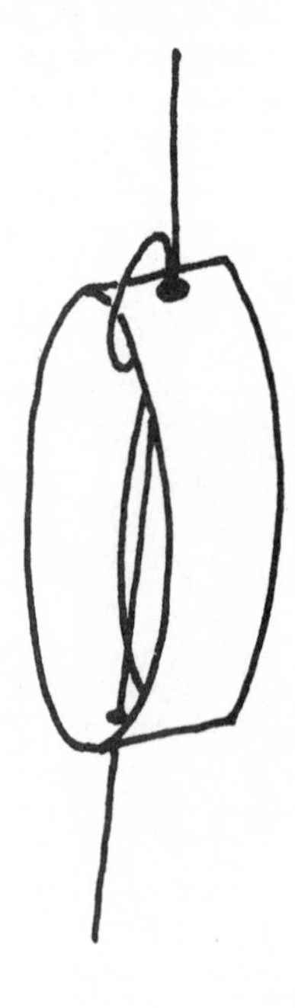

Hamburger Box Styrofoam Collector's Case

After a child enjoys a hamburger, he or she can use the styrofoam box to make a striking collector's case for many little things. Keep pins, thread, change, or whatever in a place where it is easy to locate. And it even has a clasp on the lid! A white design is very effective on this revitalized box.

MATERIALS

hamburger box (or similar)
egg cartons (white foam) or meat tray
tacky craft glue for plastic
scissors

PROCEDURE

Cut the lid off of a white styrofoam egg carton, or use a meat tray.

Then cut out flowers and leaf shapes to be arranged on the top of the hamburger box. These shapes can be overlapped (e.g., to make the centers of flowers).

Glue the shapes in place on the top of the box using the tacky white glue made for plastics.

Foil Sculpture

Used aluminum foil wrap is a very pliable medium for sculpture. Rather than discarding the foil, encourage your class to save it. It can be used to form animals, people, or interesting nonobjective shapes that will sparkle and shine like the eyes of their happy creators.

MATERIALS
used aluminum foil

PROCEDURE

First, open the foil flat. Then, pinch and squeeze it to create three-dimensional shapes that are representational or nonrepresentational. To add other pieces of foil, it is necessary to overlap the loose pieces and squeeze them together.

Alternative: Beads could be made of aluminum foil and strung on thread for a sparkling, metal-look necklace. A foil bracelet formed over a wire or cardboard ring would be worn with pride by almost any child.

Paper Things

Magazine Beads

Making paper beads is an old-time craft which your students will find interesting. It's appropriate for almost any age level that likes jewelry or stringing things. Very young children are best taught to make large beads with large holes, but older children can make very refined, sophisticated beads. The necklace shown in the photograph is made of rather large beads strung on heavy yarn.

MATERIALS

magazines with colored pictures
white glue
ruler
scissors
string or yarn

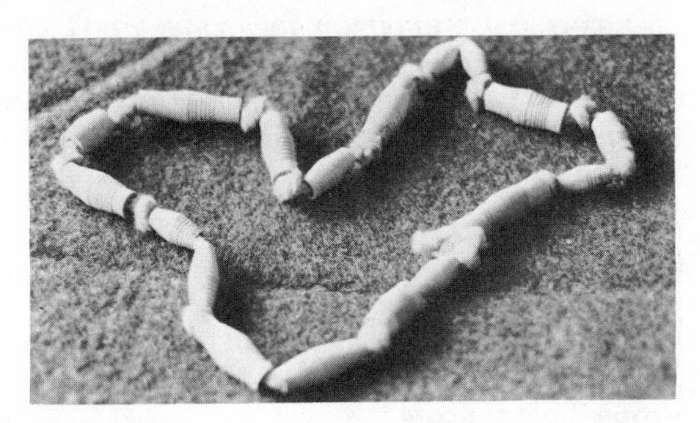

PROCEDURE

Find colored pictures in magazines. These will be cut into long, very thin triangles (see illustration). The base of the triangle will determine the length of the finished bead. The length of the triangle will determine how thick the bead will be.

For larger holes, use a pencil to roll the bead. For smaller holes, use a knitting needle, pick-up stick, or even a toothpick.

Begin rolling from the base of the triangle and keep the paper as tightly wrapped around the rolling tool as you can. Keep it tight all the way to the point of the triangle. Apply a drop of white glue to the end of the triangle and pull out the rolling tool (you may have to loosen the bead just slightly to release the tool).

Repeat the process until you have rolled all the beads that you will need for your project.

Then string them on yarn, string, leather, or whatever. A knot between each bead adds to the interest of the design.

Wear the jewelry proudly or give it as a gift.

Alternative: This technique can also be used to make curtains for room dividers, but such a curtain takes many many beads. It could, however, be done as a group project with everyone in the class working on beads for one room divider.

On-the-Dot Mosaics

Hole punchers leave holes, but what happens to the paper dots that are punched out? They probably get thrown away, which is something we can avoid.

Mosaics are a logical use for these discards. A mosaic is a composition made of small pieces of material placed next to one another. Any school department that punches a lot of paper will be happy to save the dots for you. A hand paper puncher is also easy to use to punch colors that aren't already in the assortment that is given to you. Very young children have a short attention span, and therefore may not want to see this project through to completion. For this reason, it would probably be more suitable for older children.

MATERIALS

construction paper (for the background)
paper dots
paper hole puncher
scraps of colored paper
white glue

PROCEDURE

Lightly draw a pencil design on the background paper. There are two ways of using the punches. The first is to glue each dot individually by applying glue to each one. This was the technique used in the geometric design shown.

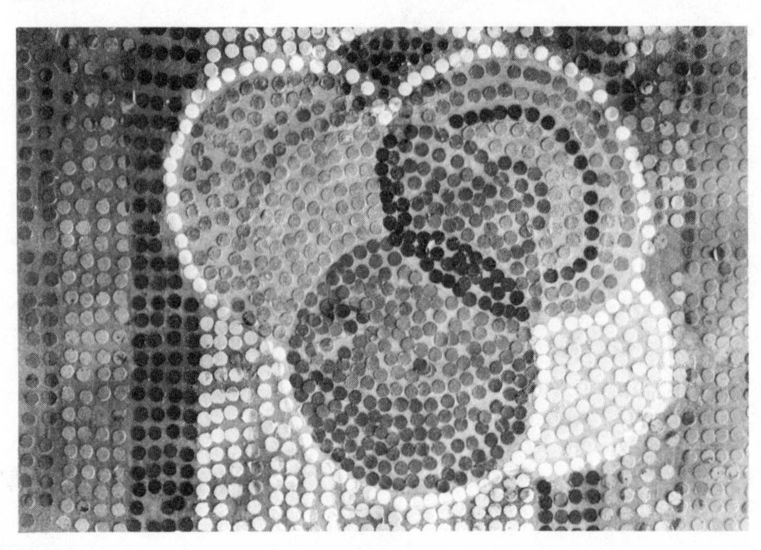

The second method, as shown in the portrait example, is to make a drawing with a squeeze bottle of white glue. Use the bottle as you might any drawing tool, allowing the glue to flow out as a white line. While the glue is still wet, sprinkle on the paper dots. (Those used in the example were rectangular punches rather than dots, but either will work the same way.) Let the glue dry completely before shaking off the excess paper.

If your punches are limited to one or two colors, the second technique works best. In this case, the entire piece of paper is not covered, as is usually true in the first technique discussed.

This project takes time, but the results are well worth the effort.

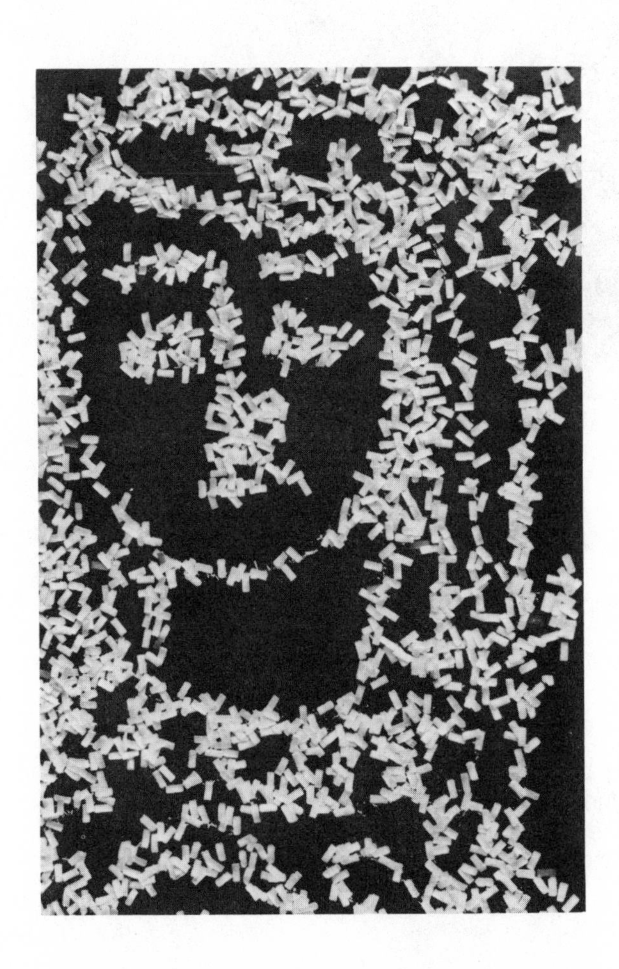

Imagination Plus

Your pupils will discover that old magazines with colored pictures provide a gold mine of material for creating photomontages or collages made of reassembled pictures. The original idea of any picture can be completely changed by using parts of it with parts of others.

Your class would never hope to see the situation shown in the pictured example. A businessman sitting on the back of a lion with a woman's head? Let's hope not! But the creation of such a scene can be a fun experience and a stimulus for the imagination.

MATERIALS

magazines with colored pictures
scissors
background cardboard or heavy paper
rubber cement

PROCEDURE

Look through magazines for interesting pictures of people, animals, landscapes, buildings, furniture, and so on. All of these may contain something you will need. At first, don't worry about which ones you will use.

As you look at a picture, you'll find that it might spark an idea to grow. Then you can start changing things around. Cut a tree out of one and some rocks out of another. A head from a girl may go on a cat. Eyes from one person may be effectively used on another—maybe even upside down!

Encourage the members of your class to move and change the various elements. When a satisfactory arrangement has been achieved, start to glue with rubber cement (it doesn't make the paper wrinkle).

Cityscape in Print

Indicate to your class that the columns on a printed page of a magazine or newspaper have a very architectural look. They vary in width. They vary in lightness and darkness, depending upon the kind of type used.

By combining columns from different sources, a whole cityscape can be created. The students will also learn about geometric shapes as the project progresses.

MATERIALS

newspapers and/or magazines
rubber cement
plain colored paper for background
scissors

PROCEDURE

Select a large piece of paper (12″ x 18″ x 24″) for the background color. This will be the color of the sky in the completed composition. It doesn't need to be blue, however.

Have the class begin cutting columns out of the newspapers and/or magazines. They should avoid pictures and concentrate on the printed words on a page. When they have a good supply of rectangular shapes, they should arrange them on the large background paper.

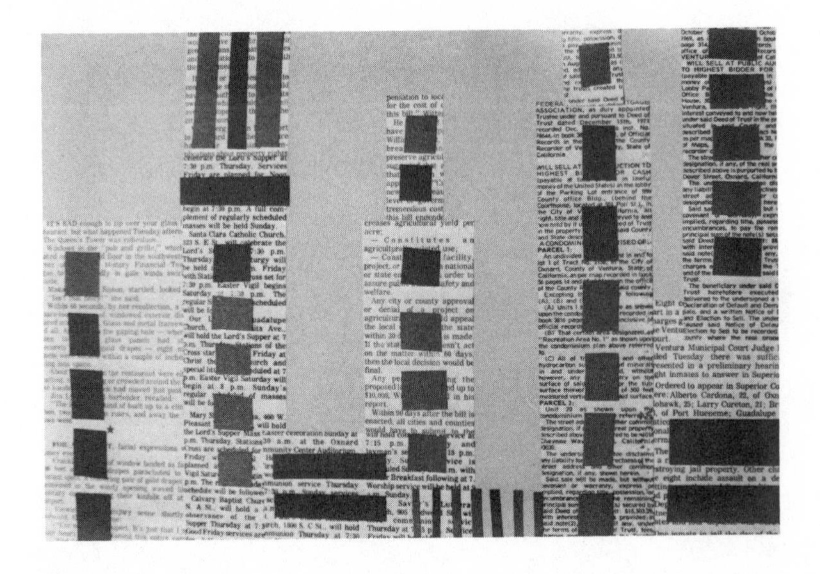

The class members should think of themselves as city planners who are going to arrange all these structures to create a skyline of buildings. They should have some skyscrapers going all the way to the top of the page, and some buildings that overlap.

After the buildings are arranged, the planners can go back and glue them down with rubber cement. Rubber cement won't wrinkle the paper as some other adhesives might.

Now add details with cut paper shapes for windows, doors, fences, lightposts, or whatever the design seems to need.

Paper Bag Thinking Caps

"Put on your thinking cap." Your students have probably heard this phrase many times. But how many times has anyone actually seen a thinking cap? Get out the scrap paper box, a paper bag, and add some imagination to make the imaginary thinking cap a reality.

MATERIALS

paper bag
scrap paper
white glue
scissors

PROCEDURE

Roll the top edge of the paper bag all the way around like a cuff. Repeat, so you have a double cuff. This will strengthen the edge and make it fit better. The size of the bag used will depend upon the age of the children, but you can always staple a pleat in it if it is too large.

The bag can be cut, folded, curled, or whatever the imagination desires to change its basic shape.

Then, have the children use colored paper scraps to design the shapes to be added tc this fanciful cap. Encourage them to be elaborate and inventive!

Luminaria Paper Bag Lanterns

If your class is planning an outdoor affair, they might consider using luminaria lanterns for decoration. These lanterns, which originated in Mexico, are often used to line the driveway to the house where a party is being held. Never *use them indoors, because the candles placed in the bags create a fire hazard.*

MATERIALS
paper bag **scissors** **sand** **candles (or pieces of candles)**

PROCEDURE

Fold the bag vertically once, twice, or more. Cut on the folds to create a repeated open design all the way around the bag. Each fold will produce an additional repeat of the design; but the more folds used, the harder it is to cut the bag (see illustration).

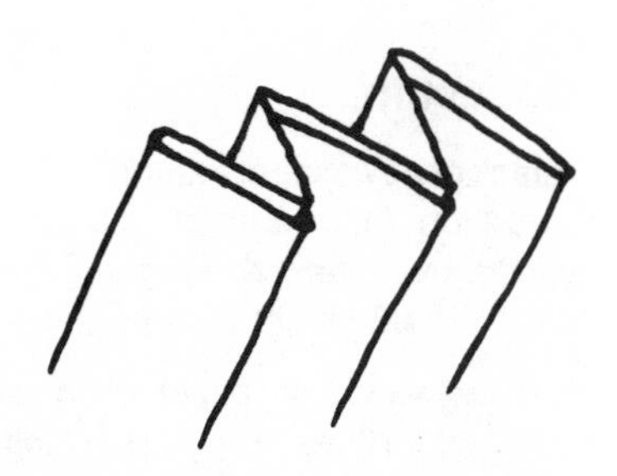

Leave an area about 4″ from the bottom uncut.

Put enough sand in the bottom to *firmly* support the candle. Set the candle as near the center as possible, so it won't tip against the bag. Partially burned candles will work fine, but short, fat candles are best. The latter will burn longer and they won't stick up too high.

Hot Crayon Banners

Some office copying machines can provide the class with paper for making see-through banners using crayons and heat. In some copiers, after the intermediate coated paper has been used, it comes out of the machine all nicely rolled. It is about 11" wide and nearly unlimited in length.

MATERIALS

crayons
intermediate coated paper from copying machine
electric iron
newspapers
wrapping paper or butcher paper
cardboard tube from a pant hanger

PROCEDURE

A stack of newspapers about ½" thick is a good surface to work on. Place a sheet of brown wrapping paper (or similar) over the newspapers. This paper seems to hold the heat longer than just newspapers.

Heat the wrapping paper by moving a very hot iron back and forth as though pressing it. When the wrapping paper is hot, lay a piece of the coated copying machine paper (the size of the desired banner) directly on top of it.

Then, begin to draw your design with crayons. The feel of drawing on hot paper is completely different from drawing on cold paper. The crayon seems to slide as the melting color is applied to the surface. When the crayon no longer melts, reheat the wrapping paper. Continue the process until the banner design is finished.

A paper stick for the top can be made from the cardboard tube on a pant hanger. If one isn't available, just roll a piece of newspaper very tightly to form a small tube. It can be rolled on a pencil for a guide. Tape the banner to this tube.

Cut or fringe the bottom, or add a second stick for weight.

Hang it near a window. The light will make the colors seem to glow.

Top It Off!

A paper plate makes an ideal base for an Easter hat or just a funny party hat. Boys as well as girls will enjoy designing headdresses for many occasions and/or celebrations.

MATERIALS

paper plate (may be used, but wash it first)
white glue
colored construction paper
scissors
yarn or ribbon
used gift box bows

PROCEDURE

Make a hole on each side of the plate and string a piece of thick yarn or ribbon through and over the top (see Illustration A).

Decorate the hat with paper flowers, feathers, and curls. Any type of paper may be used, but construction paper has good body. Glue the cut and torn shapes to the paper plate with white glue.

Illustration B shows one way of cutting flowers, or invent your own.

For additional interest, other scrap materials could be incorporated, such as buttons, foil, net, and used gift box bows.

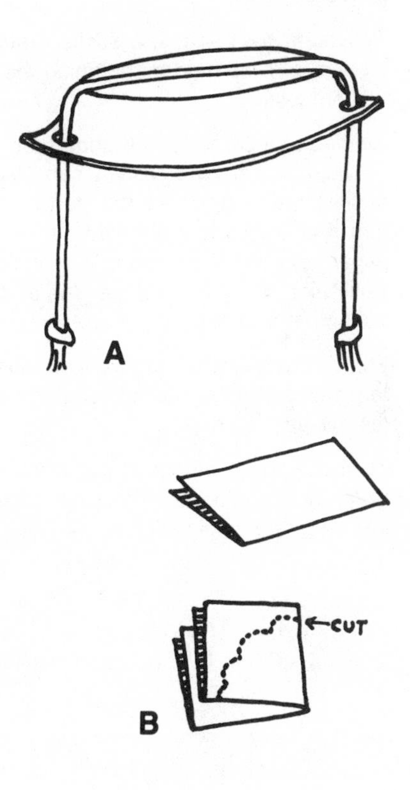

Rolled Paper Sculpture

Constructing giant structures with paper tubes will not only be an exciting project, but it will help students to develop an awareness of three-dimensional space. The possible engineers and architects of tomorrow may find motivation in this kind of activity.

MATERIALS

magazines
white glue
pencil or stick

PROCEDURE

Roll magazine pages into sticks. Get a large supply of the sticks rolled before beginning the actual construction process.

Starting at a corner of the page (use two at a time for strength), roll as tightly as you can. Use a pencil or stick of similar size to get the roll started, but remove it before it is completely covered by the paper. Hold the paper very tightly as you roll to avoid loosening (see Illustration A). A dot of white glue at the end of the roll will hold it securely.

Form a triangle, which is a very strong form, by bending the magazine stick and inserting one end into the other (see Illustration B).

To make larger triangles, put two sticks together by inserting end into end. Then bend them into triangles just as with a single stick.

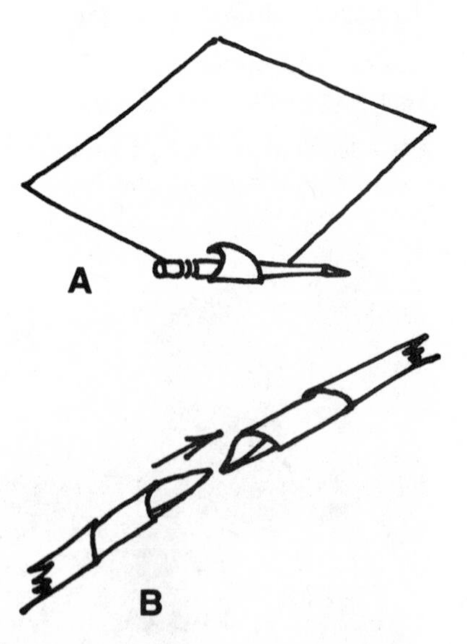

Have the children begin to build geometric constructions with the triangles. Use small amounts of white glue to attach the sticks together, holding until the glue is secure enough to support them. Be sure that each addition is supported against another piece or two. Here is where the engineering aspect comes into the project. The strength of the construction is as important as the visual design.

Size is unlimited! Two or three children working together could develop an enormous sculpture.

To the Letter

A hanging letter holder made of paper plates will keep important notices and bills easily at hand. Designed for a kitchen or a den, its convenience will be very much appreciated.

MATERIALS

two paper plates
white glue
scissors
yarn
large-eye needle or a paper puncher
permanent ink felt pens or paint

PROCEDURE

Cut one paper plate in half. Glue the face of the half plate to the face of the whole one (see Illustration A). Decorate with felt pens or paint.

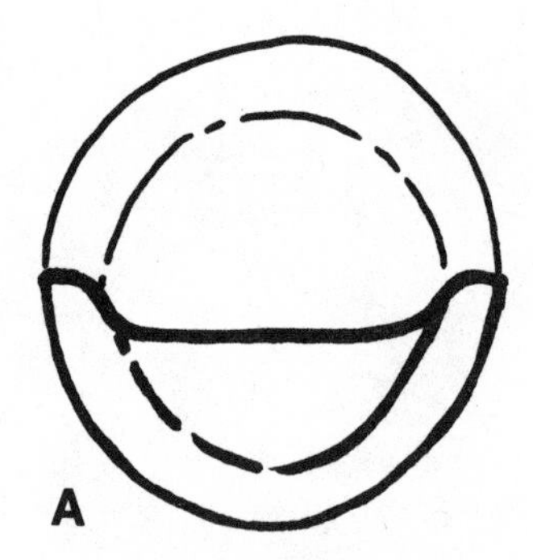

A paper puncher may be used to make holes for lacing instead of a needle (using a puncher is safer for younger children). Space holes evenly all the way around the edges of the plates. Then lace with colored yarn. Use any simple stitch, e.g., the blanket stitch (see Illustration B).

Finally, add a loop at the top for hanging on the wall. This letter holder could make a father very happy on Father's Day.

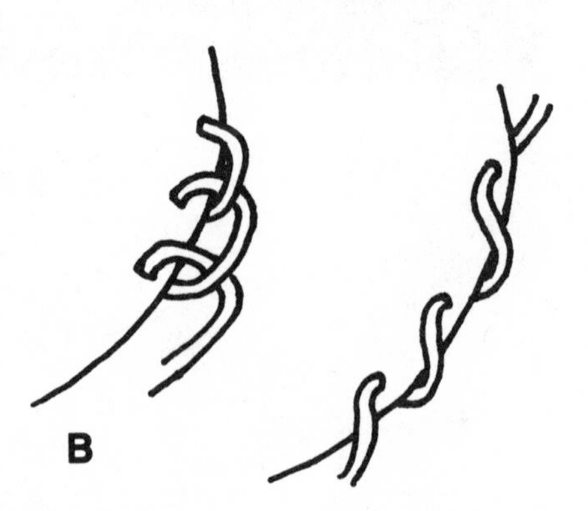

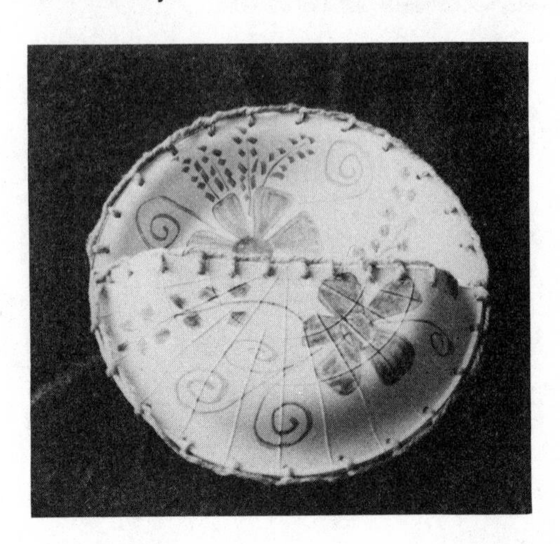

Rubbed the Right Way

A rubbing called frottage *in French is often made to reproduce the raised or carved designs on tombstones, walls, doors, or floors.*

It is an excellent way of making textures more visible and is at the same time a study in light and dark. This particular technique makes use of carbon paper that has made as many copies as it possibly can. But there is still plenty of carbon left on the surface to use for a rubbing.

MATERIALS

used carbon paper
onionskin paper or newsprint
textured materials
scrap paper
scissors

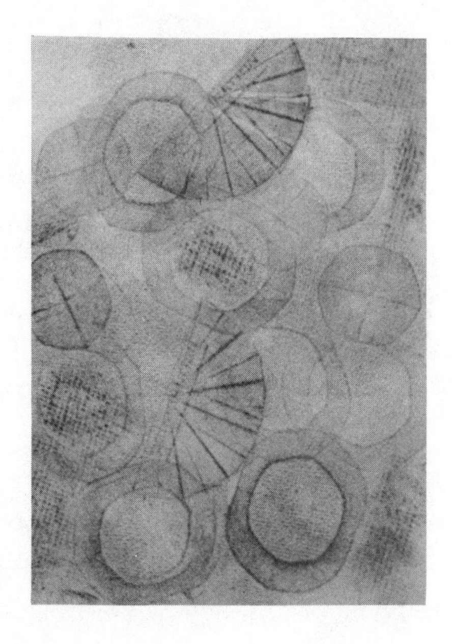

PROCEDURE

Cut shapes from scrap paper and arrange them on the table. Place onionskin, newsprint, or other lightweight paper over the pieces.

Tear the carbon paper into approximately 3″ pieces. Wad one of the small pieces up slightly and *gently* rub over the surface of the paper using a circular motion. Continue rubbing until the design is transferred onto the paper.

Move the paper shapes to make a new arrangement and repeat the process.

To introduce texture into the design, place various textured objects under the paper and rub in the same way. Don't try to make the transfer too quickly—do it slowly and gently.

Newspaper Creatures

Papier-mâché in sheet form requires very simple materials and is easy to make. Wonderful creatures can be made from these sheets by all age groups. The size of each creature and the complexity of its design will vary with the abilities of the children. But any four-legged animal, real or imaginary, is relatively easy to develop.

MATERIALS

newspapers
scissors
brushes
tempera paint
white glue (diluted 50% with water) or wheat paste

PROCEDURE

Cover the work surface with newspapers.

Dilute white glue with 50% water or mix wheat paste according to the directions on the package.

Using a bristle brush, smooth a layer of glue over a sheet of newspaper. Lay another sheet over this. Cover the second sheet with glue, and repeat the process until six or eight layers have been applied. For larger pieces, use the greater number of sheets.

When this is partially dry (feels leathery), it may be cut and formed into the desired shape.

An animal may be designed by cutting the shape as though it were a bearskin rug on the floor. Draw it with pencil first, if desired (see Illustration A).

To shape the cut piece, mold it over a tube of rolled newspapers for support as it dries (see Illustration B).

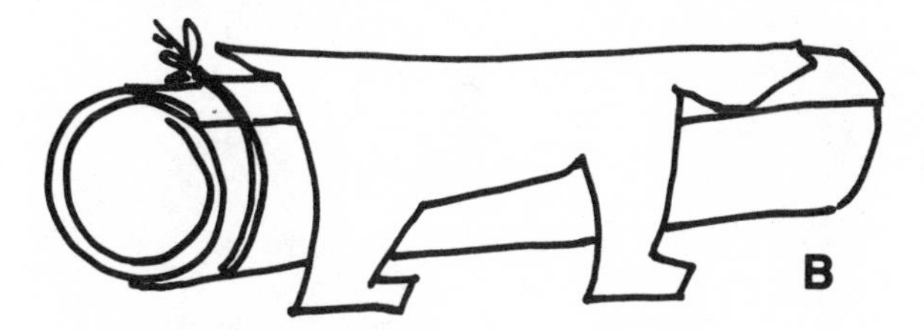

When it is dry, paint it with tempera paint. Give it stripes, dots, checks, or whatever the imagination desires.

Twist-and-Roll Strip Constructions

Your class can use tickertape-type paper strips to make birds, mobiles, and Christmas decorations. They can be made from used tape or ends of rolls. It usually comes in half-inch widths, but any width will work.

MATERIALS
paper strips **scissors** **white glue**

PROCEDURE

Roll various lengths of the strips and secure them with very small amounts of white glue (a little glue works better than a lot). Make rings and curves of different sizes (see illustration). These will be glued together to form an abstract design or one which represents a fish, bird, flower, etc.

The outside form can be constructed first and then filled in with more rolls and curls.

The completed piece might be suspended from the ceiling with a light-colored thread. It will move as the air in the room moves. Or, the shapes could be glued to a background paper as a relief plaque.

Swinging Pictures

Pictures from magazines make very interesting mobiles. Each student should choose one subject category for his or her mobile (e.g., people, furniture, flowers, bugs, and so on). The members of your class can learn a lot while searching for and selecting the pictures necessary for this project.

MATERIALS

magazines
rubber cement
scissors
needle and thread
coat hanger
colored construction paper (or similar)

PROCEDURE

Have your students search for pictures in a given subject area. Tell them to find more than they will need, so they can select a variety in different sizes and shapes.

Cut the pictures into regular shapes (circles, ovals, squares) and mount them on colored construction paper. Use rubber cement, which won't wrinkle the paper. Trim the construction paper a little larger than the picture, but follow the same shape. This will give a colored border around each picture.

Use a needle and thread to string and attach the pictures to a coat hanger. Hang them at different lengths so some will be high and some low. Any number can be assembled, even as long as from floor to ceiling!

While the finished piece can be hung against a wall, it is more effective hanging freely so it can move. It is especially interesting when it moves, since the solid color of the paper alternates with the pictures.

Magazine Mosaics

Have the children collect and save old magazines with colored pictures. They can be used to make paper mosaics. A variety of colors will make these pictures composed of many individual pieces extremely effective.

Examples of ancient Roman or Middle Eastern mosaics would be excellent motivation for the project. This project takes a long time, so the attention span of the children should be considered before beginning.

MATERIALS

magazines with colored pictures
white glue
heavy paper for the background
shellac or acrylic medium
scissors

PROCEDURE

Make a drawing on the heavy background paper (about 12″ x 18″). Keep the design rather simple, since small details are very difficult to achieve with this technique.

Decide on a general scheme. Then, look through magazines and tear out the pages that have the colors you need. Keep all of the reds together, the blues, etc. The good thing about using different kinds of magazines is that you'll have many variations of each color.

Cut the pages into small square or rectangular shapes no larger than ½", again keeping similar colors together. Small paper cups or boxes are helpful for keeping the different colors of cut paper separated.

Apply glue to a small section of the design at a time. Place pieces next to one another, thus building a solid shape from many small ones. Smaller squares may be cut to fit into very small spaces.

When the picture is complete and thoroughly dry, coat it with shellac or clear acrylic medium. This will give it a protective finish and shine, as well as help to secure the little edges of paper that might be sticking up slightly.

Wallpaper Starch Painting

Discarded wallpaper offers an interesting surface for finger painting with color and starch. Old wallpaper sample books donated by a wallpaper store or ends of wallpaper left over from a papering job will provide the material needed for this project.

The paintings can be hung as pictures or used as folder covers. Perhaps other uses will be suggested as the class works.

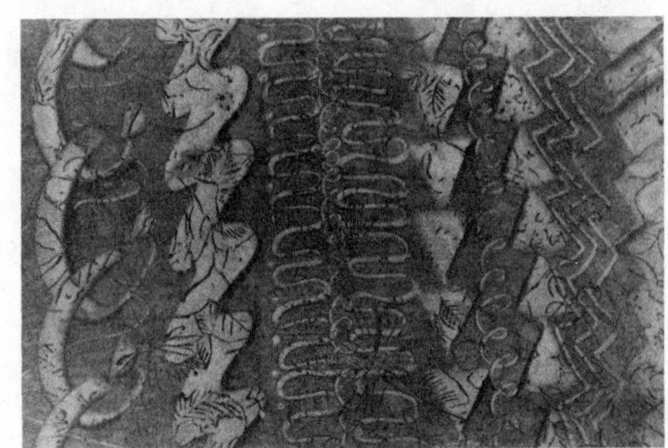

MATERIALS

wallpaper
liquid starch
powdered tempera paint
cardboard scraps
scissors
newspapers

PROCEDURE

Cut small rectangular strips of cardboard from 1″ to 2″ wide. These will be used for making designs on the painting.

Wet the table with clean water. Lay the piece of wallpaper (about 18″ x 24″) on the wet surface with the design side up. Be sure that the paper isn't prepasted. If it is, wash the paste off the back first. Otherwise you'll have it glued to the table!

Then, pour a 4″ blob of starch onto the center of the wallpaper. Smooth it out over the entire surface with the palm of your hand. If the paper isn't completely covered, add more starch.

Now, sprinkle powdered tempera paint on the starch and blend it in with the palm of your hand. When evenly distributed, you are ready to design.

Use the cardboard strips to make designs by scraping and drawing with the edges. The pattern of the wallpaper will appear as a texture in the cleaned off areas. Dry the painting on a piece of newspaper.

Mount or frame it for hanging. It also makes a good folder cover if it is sprayed or given a protective coating of shellac or acrylic medium.

Crayon Engraving on Shirt Cardboards

Shirt cardboards or the cardboards found in boxes of paper are usually discarded. These panels have a smooth, hard surface and are lightweight. So why throw away a good base for crayon engraving?

MATERIALS

cardboard
crayons
newspapers
nail, paper clip, or other pointed tool
tempera paint, acrylic paint, or India ink

PROCEDURE

Place the cardboard on a piece of newspaper. Cover it with a heavy, solid coating of light, bright colors of crayon. Press very firmly to get an even coverage. None of the cardboard should be showing when you finish. The object is to get a waxy surface.

The crayon can be applied in any kind of pattern, such as stripes, zigzag, or random.

Paint the entire surface with black, dark blue, or dark brown paint. If tempera paint is used, a little liquid soap will make it adhere to the waxy surface. Keep brushing until it is completely covered. Dry thoroughly.

Now use a sharp tool, such as a nail, an opened paper clip, or similar, to draw a design. As you scratch the paint away, the light crayon colors will contrast with the darker, painted surface. The more paint you scratch away, the more effective the design will be. Don't be satisfied with just a few scratched lines. Work for textures and light and dark patterns. The brilliant contrasts make lively drawings.

Texas Blooms

Artificial flowers made much larger than life-size are fun to use in decorating a room or hall for a party. These flowers can be made from the black and white printed sections of newspapers or from the colored comic sections. In either case, their mere size makes them very effective for children to make.

MATERIALS

newspapers
scissors
twine
tape

PROCEDURE

Each time the paper is folded you will double the number of petals that you cut at one time. For example, if it is folded twice, you'll get four petals (see Illustration A).

Use your imagination to vary the shapes of the petals and leaves. Cut several and put them together by tying them at the base (see Illustration B).

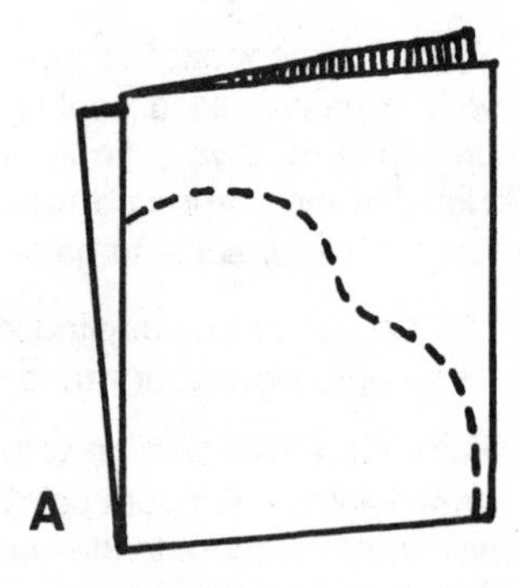

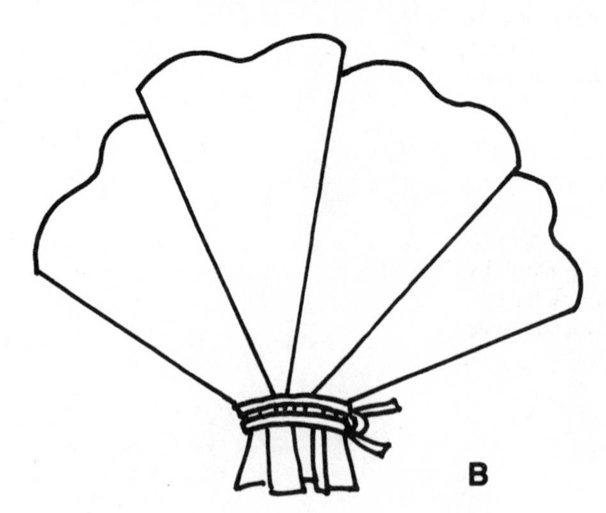

To make a stem, simply roll up a piece of newspaper as tightly as you can. Tape it at the end to keep it from unrolling. Add leaves and the flower is ready for planting in an imaginary garden.

Christmas Card Ornament

An attractive geometric form called a polyhedron (many-sided form) can be made from Christmas cards and hung as a Christmas ornament. The geometry involved will review or provide knowledge of circles, radii, and triangles. Last year's Christmas or other greeting cards can be made to have long-lasting use.

MATERIALS

used greeting cards
scissors
compass or jar lids
rubber cement
ball point pen

PROCEDURE

Collect many old Christmas or greeting cards.

Using a compass or some circular form, such as a jar lid, make a circle on the back of a card. The size will depend upon the size of the card. Very small ornaments can be made from circles as small as one inch. A four-inch circle will make an ornament about nine inches in diameter.

Cut out the circle and use it as a pattern to make 21 more circles from the designed sides of the cards. Cut very precisely to make sure they will fit together when assembling.

Now, fold one circle in half, and in half again. This will give the radius of the circle (see Illustration A).

Using this radius as your guide, make marks on the edge of the circle all the way around. Measure exactly along the edge (see Illustration B).

Then, connect every other mark with a straight line to create a triangle (see Illustration B). Cut out the triangle.

Using the triangle as a pattern, draw a triangle on the back of the other 20 circles with a ball point pen. The pen will make scored lines to make the folding easier if used with pressure. Then, fold all circles on the lines to make a triangle within each one.

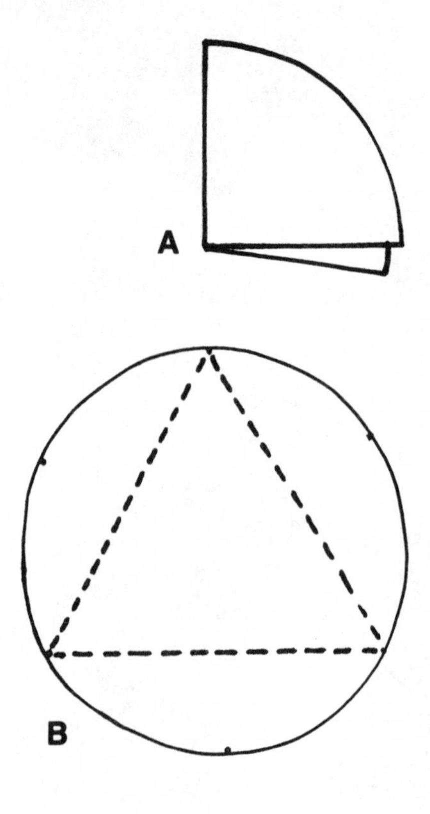

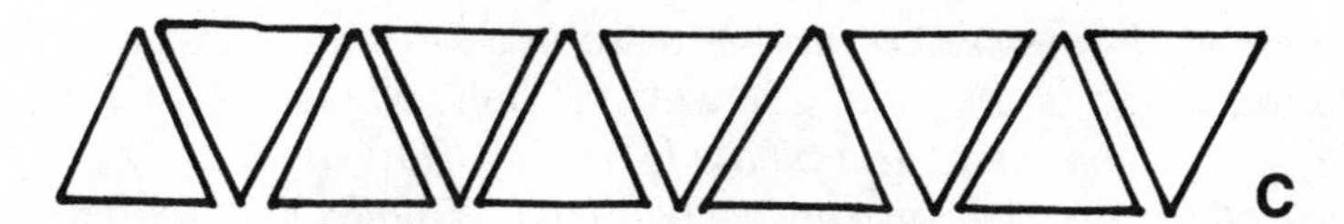

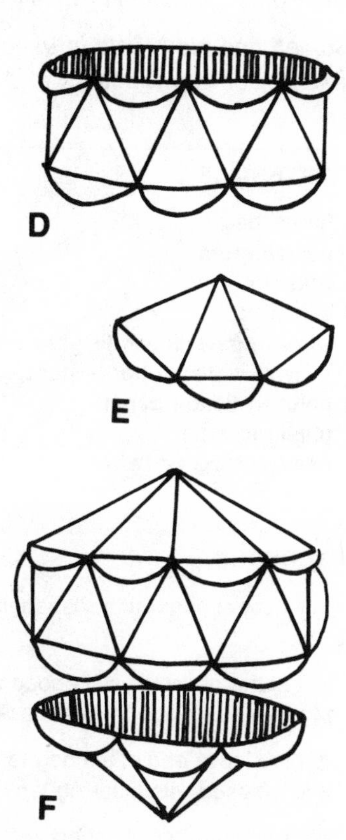

Line up the triangles of ten circles (see Illustration C). The points of the triangles will be alternated up and down. Glue them together with rubber cement. This will give you a row of these shapes. The ones on the ends of the row should then be glued together to form a ring (see Illustration D).

Glue five circles together with the points of the triangles meeting at the top. This will look like a beanie (see Illustration E).

Repeat, so you have two beanies.

Glue one of the beanies on the top of the ring, and the other on the bottom—completing the ball shape (see Illustration F).

Hang them in groups of different sizes and let the air in the room move them gently. Or, they could be hung on the Christmas tree as ornaments.

Paper Bag Piñata

The piñata comes to our country from Mexico and is a tradition at many Mexican celebrations or fiestas. The real origin of the piñata, however is Italy, where they were made with a decorated clay pot as the base. The piñata is usually filled with candy or small gifts for the children to gather when it is broken in a blind man's bluff sort of game.

Sometimes the piñata is used purely for decoration, and therefore it is not destroyed. This paper bag piñata is of that type and makes a good group project.

MATERIALS

paper bag
newspapers
brushes
scissors
wheat paste or white glue diluted 50% with water (latter is best)
colored tissue paper
masking tape
heavy string or twine

PROCEDURE

Stuff a paper bag with newspapers wadded up to give it body.

Tie it at the opening, with enough loose bag at the end to form a head (see Illustration A).

Stuff the loose end of the bag to form the head shape. Tape it closed with masking tape or wrap it with string.

Tear (don't cut) newspapers into strips about 1" wide. Using a brush, cover the strips with glue. Then, wrap the strips around the stuffed bag. Make the strips overlap in several directions for strength. Smooth them down by brushing over them with additional glue. Apply three or four complete coatings. At some time during these coatings, wrap a string around it several times, leaving a long piece at the top for hanging.

The covering procedure may take several days. It can dry between coatings, but it isn't necessary. Hanging will allow air to circulate around the whole piece for even drying.

When completely dry, the piñata is ready for decorating with colored tissue paper (or it can simply be painted with tempera).

Tissue Covering. Cut 4″ to 6″ wide strips of tissue paper as long as the sheets will allow. Fold them in half lengthwise. Cut slits all the way along the folded edge to create a fringed effect (see Illustration B).

B

Unfold and fold back in the opposite direction very loosely—without creasing.

Glue these in rows over the entire piñata, making a very fluffy surface (see Illustration C).

Cut tissue paper or construction paper details, such as eyes, beak, and tail, and glue them on for additional decoration.

Hang it and enjoy it.

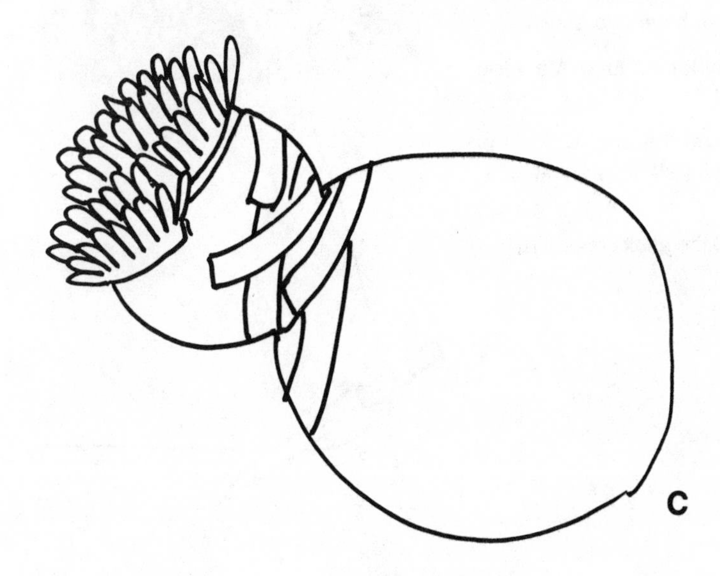

C

Flying the News

When the March winds begin to blow, it's time to think of kite flying. Point out to your class that while going out to the store and buying a kite may seem to be the easiest thing to do, it is not nearly as much fun as making one's own out of yesterday's newspaper.

MATERIALS

newspapers
tempera paint
white glue
brushes
kite string
scrap fabrics

PROCEDURE

With two full-size sheets of newspaper, one on top of the other, roll a stick as tightly as you can. Begin at the corner and roll diagonally, holding it firmly as you roll. A pencil will help you get the roll started (see Illustration A). Repeat the process so you'll have two sticks.

Cut one stick 24″ long and the other 20″ long. Mark the middle of the shorter stick (at 10″).

Mark the long stick about 8″ from the end. Cross the sticks at the marks and tie them together by crisscrossing the strings over them.

Make ½″ slits on all four ends of the sticks (see Illustration B).

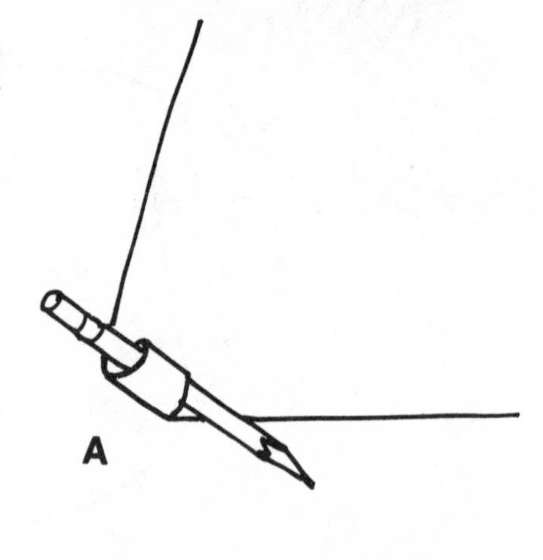

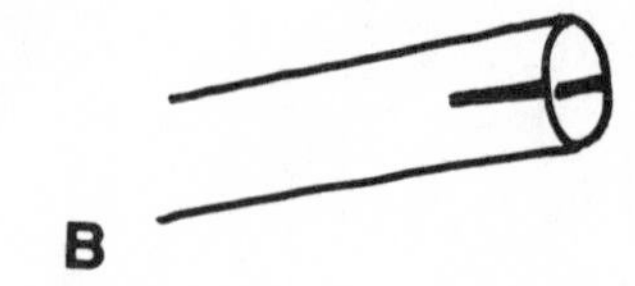

Run a string from one stick to the next by going through the slits. Keep it tight (see Illustration C).

Lay the strung sticks on two layers of newspaper and draw a straight line about 1½″ from the string all the way around. Then cut along the lines.

Trim off the points so the ends just meet the sticks (see Illustration D).

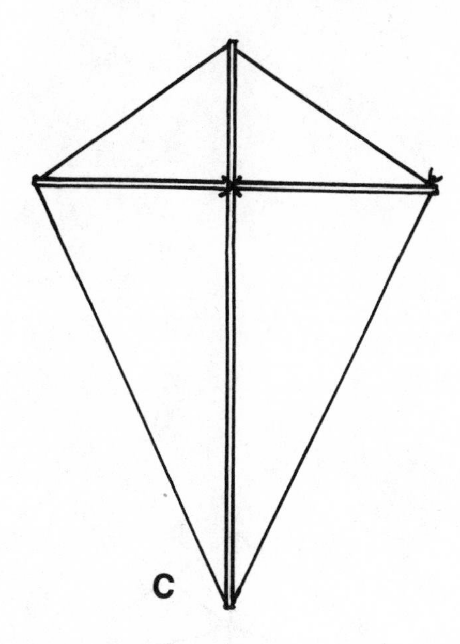

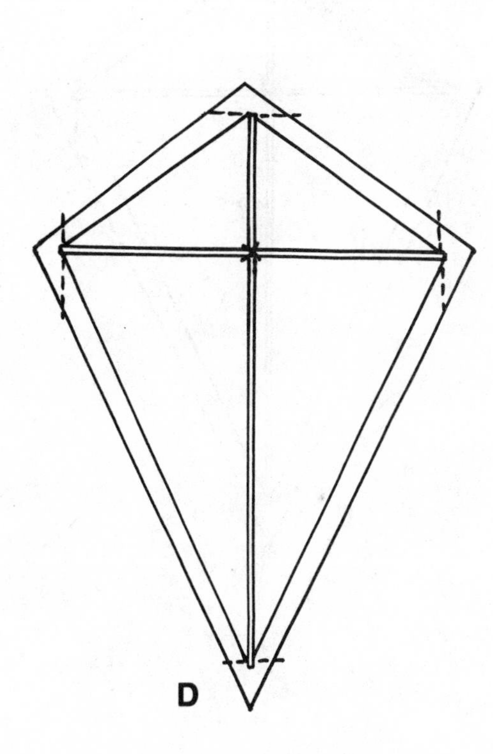

Flying the News

Fold and glue the newspapers over the string on all four sides (see Illustration E).

Paint a design, such as flowers, insects, or a face, on the front of the kite with tempera paint. Allow to dry.

Bridle. Tie strings from end to end on the short stick on the front side of the kite, also from end to end on the long stick. Leave enough slack to keep the strings about 5″ or 6″ away from the kite (see Illustration F).

Tie the flying cord to the middle of the two bridle strings.

Tail. To stabilize flight, add a 6′ tail made of scraps of fabric. Simply tie narrow strips together with knots along the way. Tie the tail to the bottom of the kite.

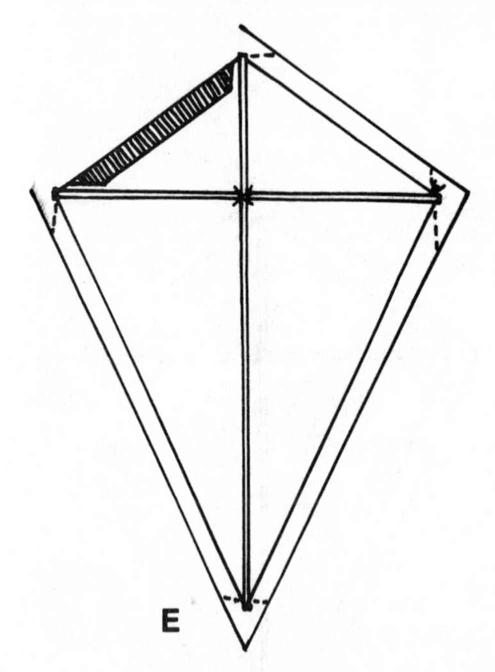

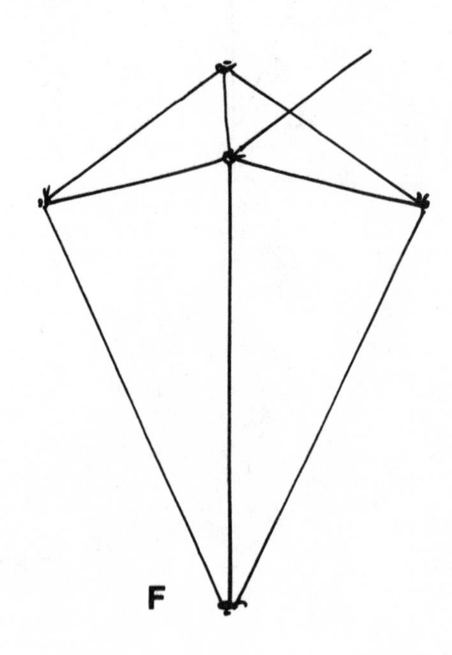

Bags That Talk

The paper bag puppet has been a favorite for a long time. Making and using one is a delightful experience for children of all ages. Lunch bags, because they are a good size for the hand to fit into, are probably best for this project.

MATERIALS

paper bag
crayons
paint
yarn
scissors
glue

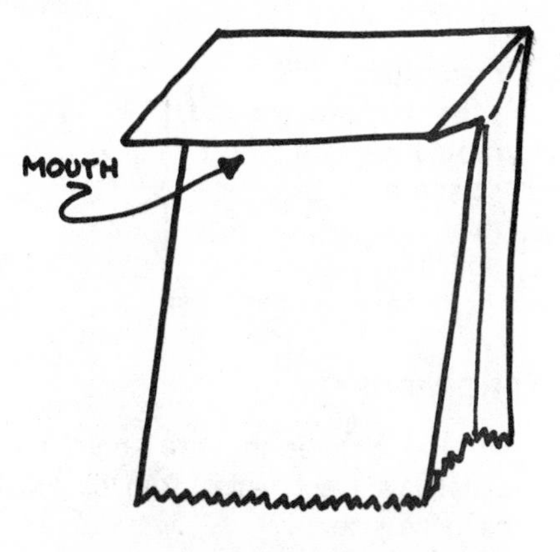

PROCEDURE

The bottom of the bag will become the mouth (see illustration).

From this point on, the door is completely open to the individual's preference, and is limited only by the availability of materials.

The character of the puppet can be developed with any of the materials mentioned. Add eyes, nose, ears, hair, and whatever else is needed to make the puppet.

To operate the puppet, the child puts his or her hand inside, keeping the fingers all the way at the edges of the bag. He or she can then move the mouth with fingers and thumb to make the puppet tell a story, recite a poem, or crack the latest joke.

Magazine Double Look

This project uses an odd visual effect for an unusual double view. First, class members must search in magazines for two related pictures. Assembling them on an accordion-folded paper creates a double look, as seen from two different views.

The project requires careful planning and measuring, which is an important learning experience.

MATERIALS

magazines
paper for background
rubber cement
scissors
ruler
pencil

PROCEDURE

Find two pictures that are about the same size and related in subject matter. Trim them so they are exactly the same size.

Accordion-fold a piece of paper that is the same height as the pictures and twice as wide. Make the folded sections about 1" to 2" wide (see illustration).

Measure the width of the folds. Then measure and cut the pictures into strips that will fit the folds on the paper. Keep the pictures in order as you cut them.

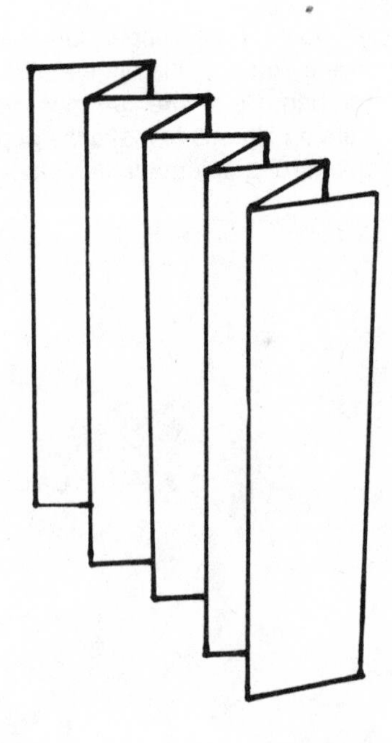

Glue the strips of each picture in sequence on alternating sections of the accordion-folded paper.

Stand it up or hang it on the wall. You'll see one picture from the left side and the other picture from the right side.

Ins and Outs of Paper Weaving

Op design, or creating an experience for the eye to explore and enjoy, can be accomplished by weaving strips of old newspapers into a piece of plain paper.

This is an extremely simple technique, yet one that will help children not only learn about weaving, but also about symmetry.

Make a place mat or wall decoration.

MATERIALS

newspapers
white glue
scissors
ruler
construction paper

PROCEDURE

Fold the construction paper in half. Draw a straight line about 1″ from the open edge. Cut slits *from the fold* to the line drawn (see illustration).

Open the paper flat.

Since the cuts were made from the folded edge, it will be a symmetrical design when it is opened.

Cut newspapers or magazines into strips ¼″, ½″, or 1″ wide.

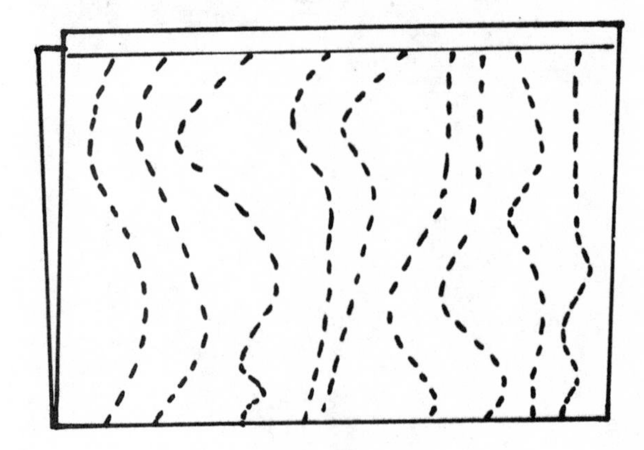

Ins and Outs of Paper Weaving

Weave the strips in and out of the slits in the paper. The second strip will go over and under in the opposite way from the first strip. The third will go over and under in the same way as the first, and so on.

Push the newspaper strips as tightly together as you can during the weaving process. This will create a heavy, solid mat.

When the weaving is complete, trim the strips to the same length as the paper into which they are woven.

A dot of glue under each strip will hold it down securely.

Mask à la Grocer

A giant mask is easily designed from a large grocery bag. The very large size bag will make a mask that covers not just the face and head, but the entire upper part of the body of a small child. Use these for Halloween or for doing exotic dances.

MATERIALS
grocery bag
scissors
crayons or paint

PROCEDURE

Put the bag over the head. If it covers the shoulders, sections can be cut out on the sides to allow arm movement (see illustration). Or, if the bag is large enough, it can go all the way down as a full body covering.

Another approach is to cut a chin shape and trim the sides and back off, as shown in the photograph.

With the bag on the head, feel for the eye placement and mark with crayon. Then cut holes with the bag off the head (they don't have to be where the eyes on the mask are located). Make the holes large enough to see through.

Draw or paint a face. Other materials can be added, if desired.

Picture Cube

This is a cube, or block, with pictures on all sides. Your class can make a number of these three-dimensional photomontages with pictures from old magazines. Older children will also learn about geometry while making the cube, but children of all ages will find it an interesting project.

MATERIALS

magazines
white glue (diluted 50% with water)
tagboard or similar cardboard
scissors
brushes
ruler
ball point pen

PROCEDURE

Decide on the size the cube will be so you can determine the tagboard size needed. The cube is made from three squares in one direction and five in the other direction. Therefore, if the cube is to be 3″, the tagboard will have to be 3 x 3 or 9″, by 3 x 5 or 15″.

For a 2″ cube, it would have to be 2 x 3 or 6″, by 2 x 5 or 10″.

Cut the tagboard the appropriate size. Mark it with squares (see illustration). Use a ball point pen and ruler to make the marks heavy enough to make scored groves in the tagboard. This will make folding easier.

Cut on the heavy *dark* lines and fold all of the others. This will create a cube. Glue all of the loose flaps with undiluted glue.

Select pictures from magazines in any particular subject area (e.g., birds, people, flowers, and so on).

Cut them out and glue them to all sides of the cube, overlapping to cover the tagboard. Brush the surface of each picture with diluted white glue to be sure it is securely adhered to the cube.

Hang the cube from a string or put a group of them together on a table like building blocks.

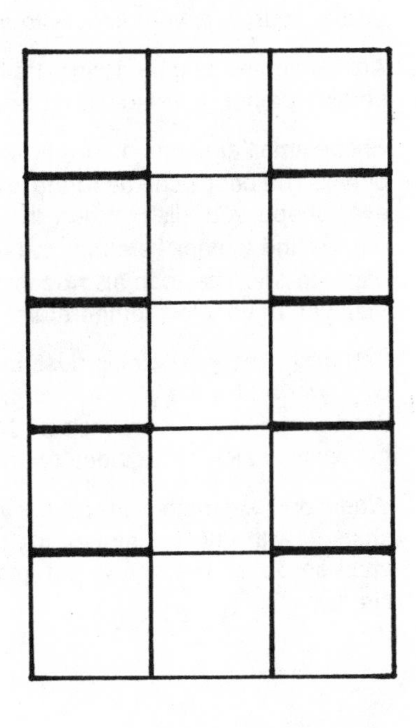

Mulch Papier-Mâché Beads

These beads, which can be beautifully colored, are made of papier-mâché and can be strung as necklaces to be worn or given as gifts. Such necklaces can adorn the necks of ladies of all ages. Though they are made from newspapers, their value lies in their unique handcrafted quality.

MATERIALS

newspapers
white glue (diluted with 50% water)
bowl or bucket
straws
tempera paint or acrylic paint
round toothpicks or nails
waxed paper

PROCEDURE

Tear newspaper into small pieces no larger than 1" square. Put them into a bowl or bucket and cover with hot water. The quantity will decrease as it soaks so be generous. Let it stand overnight.

Squeeze out all of the water that you can. Work it by pulling apart with your fingers to make it mushy.

Add diluted white glue. Using the hands, mix until it all sticks together. It will look and feel like clay.

Shape small amounts of the mulch around a toothpick or nail. The beads can be round, oval, square, or whatever shape you wish. When the beads are formed, remove the toothpicks and place the beads on waxed paper to dry. They can also be force dried on a pan in the oven at very low temperature.

Painting. Acrylic paint is best, but tempera paint will do if you add white glue for permanence. Paint each bead with one or several colors. They can be dried on toothpicks stuck into a piece of styrofoam.

When dry, string the beads on yarn or string interspersed with drinking straws that have been cut into short sections. The straws can also be painted if desired.

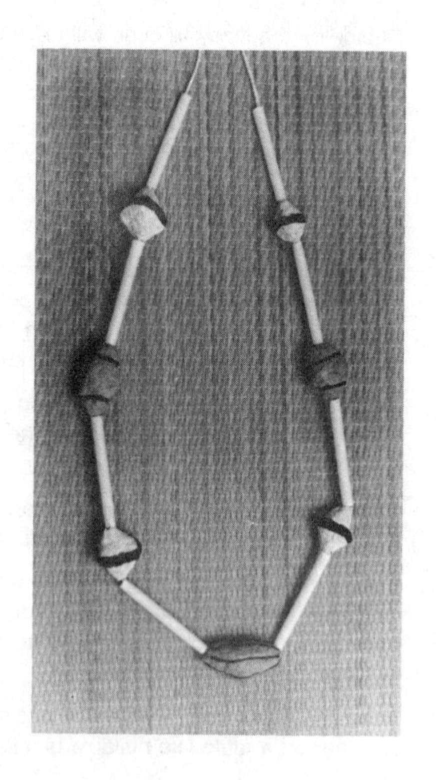

Tiles & Tops & Wires & Nails

Nails in Relief

Many heads are better than one, particularly in these decorative relief sculptures. Children love to drive nails into wood. It's especially good for developing coordination in younger children, but all ages will feel a sense of accomplishment in this project.

Nails come with many different size heads as well as colors. A piece of scrap wood is needed for the base–and ear plugs would help the instructor!

MATERIALS

variety of nails (2″ or shorter)
scrap wood (at least ¾″ thick)
hammer
newspapers

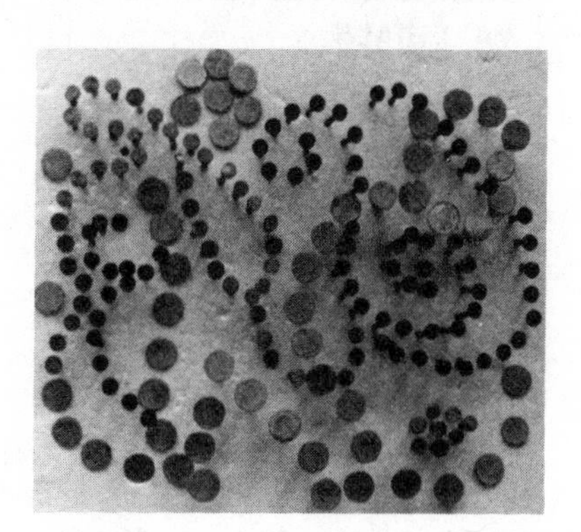

PROCEDURE

A design can be drawn lightly in pencil on the wood. This is not necessary, but might be easier for some children.

Place the wood on a stack of newspapers, which will help to absorb the sound of the pounding. It is also better to work on the floor or outside.

Begin driving nails into the wood to create a design from the nailheads. Place them close together so they appear as dots of various sizes.

Some of the nails can be driven in farther than others to create an added dimension in the relief, but try to avoid going all the way through the board. The thicker the wood, the less chance of going all the way through.

When finished, it can be hung as a relief plaque. A ring from a soft drink can securely stapled to the back will serve as a hanger.

This is a good Father's Day gift.

Painting with Nails

Pointillism is a system of creating pictures of tiny dots. It is one of the outgrowths of the impressionist style of painting. A pointillist picture is easily painted with a nailhead as the painting tool. A reproduction of a Seurat painting will provide good motivation for the class, even though this is actually a printing method, since the paint is "stamped on" rather than brushed.

MATERIALS

tempera paint
TV dinner tray (frozen type)
paper
assorted nails
newspapers

PROCEDURE

Put a small amount of paint of several colors into an aluminum TV dinner tray. This is another discard that serves well as a paint palette.

Lay the paper on a stack of newspapers to give a soft, cushioned printing surface. A light pencil sketch may be made, but it isn't necessary.

Dip the head of the nail into the paint and stamp in onto the paper. Long nails are easier to handle than shorter ones. Place the dots of color close to one another and gradually build a composition.

You'll find that when colors are placed close together, the eye will optically blend them when they are seen from a distance.

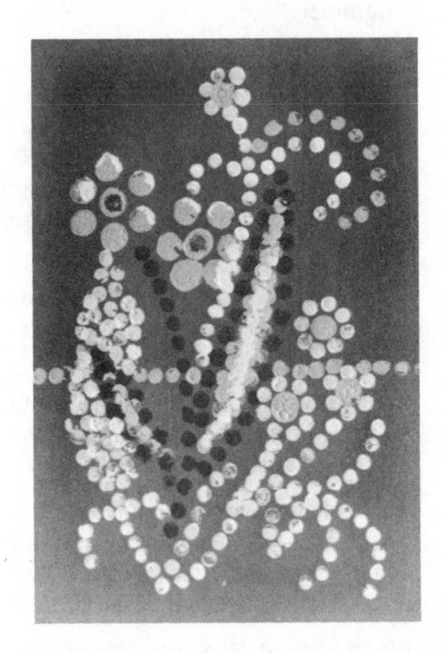

Line in Space

A wire coat hanger provides the strength necessary to stretch string or thread in space. Stretching string or thread across a coat hanger is like drawing with line in three dimensions rather than on a flat surface. When finished, these can be hung as mobiles or set on a table as sculptures.

MATERIALS

wire coat hanger (without cardboard)
white glue
string or thread (one or several colors)
scissors

PROCEDURE

Bend the coat hanger into a simple shape. Keep it simple, but going in more than one direction. You want to create a rather closed three-dimensional form.

Tie a piece of string or thread to the hanger and begin working lines back and forth from wire to wire. Wrap the string around the wire twice at each point—keeping it taut, but not so taut that the other strings sag. Watch the whole design as you work.

When it is necessary to add more string, begin where you ended by gluing the ends together with white glue.

As a final step, apply a drop of white glue to *each* spot where the string goes around the wire.

Now the piece is ready for display. Set it on a table or hang it—but enjoy it.

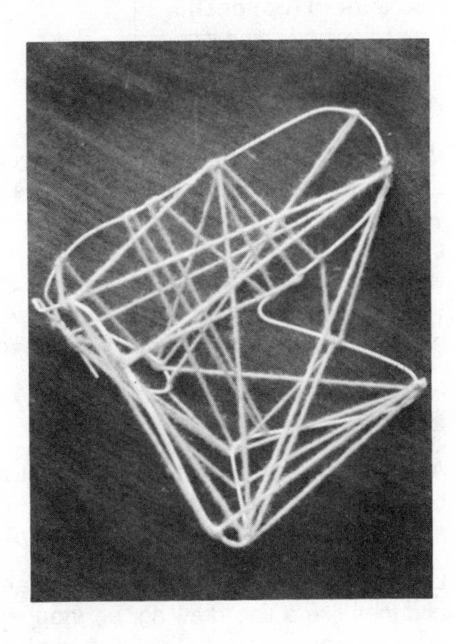

Sculpture with a Twist

Wherever you can find electrical wiring or phones being installed, you will be able to find supplies for making these wire sculptures. The scraps from these installations are plastic-coated wires that can be cut and bent easily, so little fingers can manage them. These twist-sculptures are an interesting and exciting exercise in three-dimensional design.

MATERIALS

wire scraps
scissors
wood blocks (optional)
cardboard (optional)

PROCEDURE

Often, the wire you'll find is composed of several strands twisted together or may even be part of a larger wire with a plastic coating. If such is the case, you should separate it so you have single strands to work with. Experiment with the flexibility of the wire to see what the fingers can do with it. Most wires that you find can be cut with scissors. If not, you'll need a wire cutter.

Pieces may be twisted together to make them longer or to add new shapes.

Animals, flowers, insects, or nonobjective shapes can be made as free-standing sculptures or they can be stapled to a block of wood or a piece of cardboard.

Let your imagination go! Your design can be very open and look like a line drawing, or wrapped many many times to make an almost solid shape.

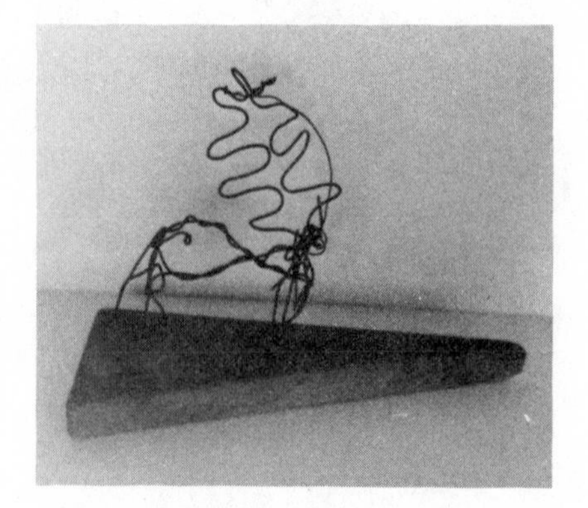

Decorative Coat Hanger

A coat hanger covered with a beautiful yarn design is not only decorative, but will also prevent clothing from sliding off onto the closet floor. It can be made using simple, square macramé knots, and provides a way to put scrap ends of yarn to good use (for function and design).

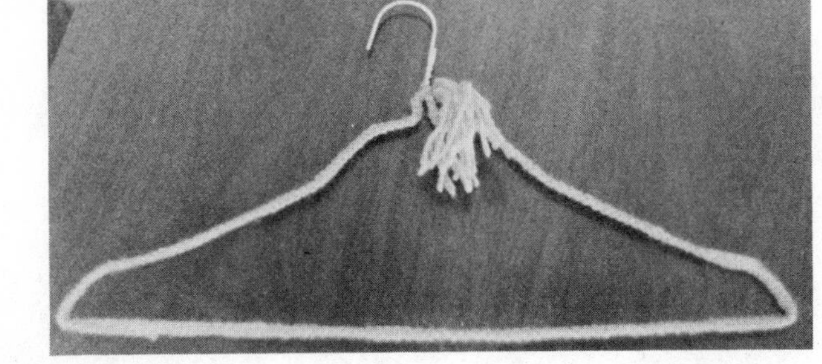

MATERIALS

wire coat hanger
yarn scraps
white glue

PROCEDURE

Find a sturdy coat hanger, preferably without the cardboard tube on it.

Determine the middle of the yarn strand. Work with pieces no longer than 72″ for ease of handling. In this case the middle would be at the 36″ mark. Tie the yarn to the hanger (at the twisted part near the hook) at the middle of the strand.

Begin making square knots and continue all the way around until they meet at the point where you started. When you run out of yarn, simply glue the end to the hanger and begin in the same way with a new strand.

Square Knot. To tie the first half of the square knot, place the right-hand string over the hanger and hold it there. Now take the left-hand string and place it over the first string, go under the hanger and up through the open space between the first string and the hanger. Pull on both strings to tighten the knot (see Illustration A).

To tie the second half of the knot, reverse the operation. Place the left-hand string over the hanger and hold it there. Then take the right-hand string and place it over the first string and under the hanger and up through the space between the first string and the hanger. Pull on both strings to tighten the knot (see Illustration B).

Be sure to begin the knot from alternate sides to make a complete square knot. If it begins to twist, you are probably starting from the wrong side. If this happens, take it out and do it again.

Add a tassel, a bow, or beads for added decoration.

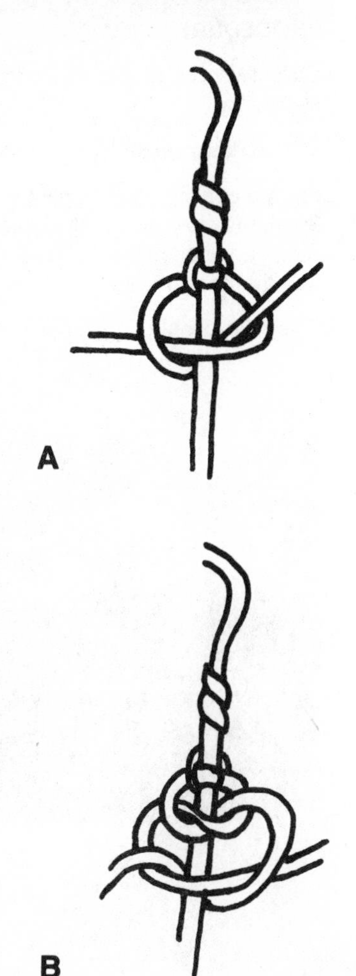

Bits and Pieces

Linoleum tile scraps can be broken very easily with the fingers, so they are a good material for your class to use for a mosaic (a picture composed of many small pieces). Ceramic tile can also be used, but a tile "nipper" or hammer is required to break it.

Tile can be found in all colors, but if only a limited color selection is available, some of the scraps may be painted with enamel.

MATERIALS

linoleum tile scraps
white glue
heavy cardboard

PROCEDURE

Collect scraps of tile. Paint them with enamel if it is necessary.

Draw a design in pencil on a piece of heavy cardboard.

Apply a coating of white glue to a small area at a time. Break the tiles into small shapes and place them next to one another on the wet glue. Continue in this manner until the design is completed.

When thoroughly dry, the picture is ready for hanging.

Paper Cup Lid Faces

The lids from paper cups of all sizes provide good round shapes to turn into faces. These amusing faces are suitable for mounting or hanging from strings as mobiles.

They might also be used on a Christmas tree to represent all of the members of a group or family. The tabs provide a place to put the hooks.

MATERIALS
paper cup lids
crayons or paint

PROCEDURE

Wash and dry the lids.

Draw faces on them with crayons or paint.

What could be easier?

Jingle Tops

Metal bottle caps make a jingling sound when they are strung together. We can therefore use such caps to make a rhythm instrument similar to one known as a sistrum, which is used in religious rites.

Use these jingling instruments to keep time to music in a rhythm band or as an accompaniment to improvised dances.

Young children can make these with assistance, but they must be careful when handling the sharp metal caps.

MATERIALS

6 to 10 metal bottle caps (not the twist-off type)
pant hanger
rubber bands
hammer
large nail
scrap of wood
pliers

PROCEDURE

Place the bottle cap on a scrap of wood and make a hole in the center with a large nail and hammer. Handle *carefully*, since the hole will be sharp.

Take the cardboard roll off a pants hanger (save it for another project). Using a pair of pliers, bend the ends of the hanger into a closed loop (see Illustration A).

Squeeze the hook of the hanger to make a handle.

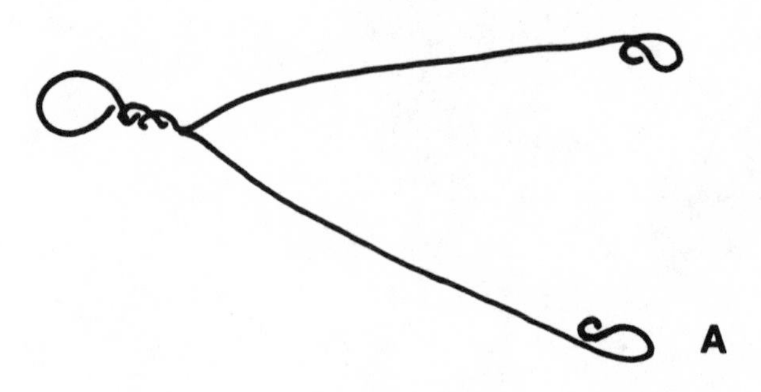

Slip a rubber band into one of the loops (see Illustration B).

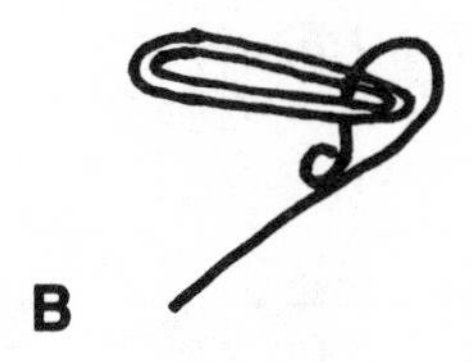

String the rubber band through the holes in the bottle caps, alternating their direction. You may need a toothpick or something similar to help get the rubber band through the holes. However, be careful not to poke a hole in the rubber band since this will weaken it.

When all of the caps are strung, slip the other end of the rubber band into the loop on the other side of the hanger.

The tension of the hanger will stretch the rubber band. As you shake it, the caps will jingle together.

Option: Caps could be painted with enamel before stringing.

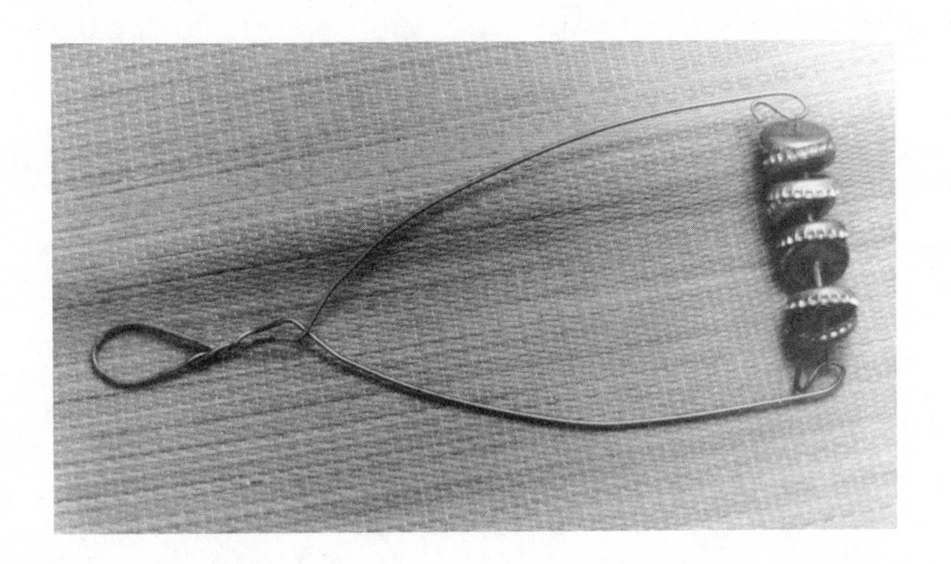

Ring-a-Ding Wind Chimes

The kitchen trash basket may hold the metal material needed to make a wind chime. Every time you open a vegetable or soup can, save the lid you cut out and also cut a second disc from the bottom. When you have many such metal discs, they can be used to make a decorative wind chime which will make lovely sounds as it sways in the breeze.

MATERIALS

2 lb. or 3 lb. coffee can lid
discs cut from many tin cans
toothpicks
monofilament nylon thread (fishing line)
hammer
large nail
pieces of scrap wood

PROCEDURE

Place a large coffee can lid on a piece of scrap wood and make nail holes all around the edge. Keep them from ½″ to 1″ apart, depending upon the number of discs there will be on the completed chime.

Make one nail hole on the edge of each of the other lids.

Thread a piece of monofilament nylon thread down through a hole in the coffee can lid. Go across and up through another. Then, go back down through the next hole. Now, go back across and up through the hole next to the first one (see illustration).

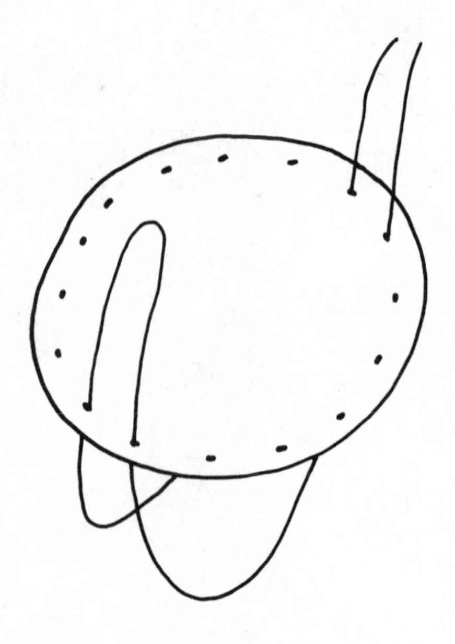

Tie all four strands in one knot for hanging.

Hang it in a doorway or in a place where it can hang free while you work on it. Tie nylon strings to small pieces of toothpicks. Thread the other end down through a hole and tie on a lid. The toothpicks will keep the thread from coming through. Continue adding more toothpicks and more lids until the chime is finished. Vary the length of the strings so some lids will hang high and some low, but be sure they will touch one another.

It also helps to alternate sides when adding the lids in order to keep the chime balanced.

Hang the finished chime in a protected area where it can move in a gentle breeze and produce its musical sounds.

Tile Memo Boards

Your pupils can make memo boards from acoustical tiles. These tiles sometimes need replacing or get chipped during installation. Don't throw old or damaged tiles away! Your students can turn them into decorative and functional little message holders.

MATERIALS

acoustical tile
straight pins
white glue
scrap fabric
scissors

PROCEDURE

Cut a piece of fabric about 6″ larger than the tile so there will be enough to wrap around the back.

Lay the tile on the middle of the wrong side of the fabric. Wrap the fabric around to the back of the tile, pulling it as tightly as you can. Push straight pins into the tile at an angle to secure the fabric.

Turn the tile over.

Cut fabric shapes out of scrap fabric to decorate the corners or edges of the front. Glue the pieces on with white glue, but be sure to leave enough blank space for pinning notes.

Going in Circles

Plastic tops from bottles can be used to create an interesting, repeated pattern for a wall relief. It's an activity which will help your students develop an awareness of pattern and variety in sameness. Even though the shapes are the same, changes in color and decoration will produce differences in this simple design.

MATERIALS

plastic bottle tops
white glue
tempera or acrylic paint
cardboard (very heavy)
brushes

PROCEDURE

Arrange the bottle tops on a piece of cardboard. Place some upside down and some right side up. When they are arranged to fit the area, glue them down with white glue. The cardboard may have to be trimmed so the tops will go to the edges.

Use tempera paint or acrylic paint to add color to the circular shapes. Paint some of them on the inside, some on the top, and some on the rim. A lot of variation can be achieved with the otherwise monotonous arrangement.

Tape a pull ring from a drink can on the back for hanging.

Linoleum Tile Printer

Linoleum printing is a craft that has been enjoyed for years in many parts of the world. Linoleum designs can be printed on paper or on fabric. You need not buy any linoleum if scraps are available. Your pupils can use small, cut, or broken pieces of linoleum tile. Because of sharp tools needed, it is recommended that this project be reserved for older children under good supervision.

MATERIALS

linoleum tile
craft knife or linoleum cutting tools (e.g., X-acto)
brayer
printing ink
lightweight paper (e.g., newsprint)
newspapers
styrofoam meat tray or extra piece of tile

PROCEDURE

If the tile being used is irregularly broken or cut, you might want to trim it to a rectangle or a square. However, this is not necessary, since the irregular shape could be used for printing as well.

Pupils may draw the design plan on the tile first. Remember that the parts of the design that are cut away will *not* print, but will remain the color of the paper being printed on. Also, the design will print in reverse, so words or numbers need to be backwards on the tile design.

To cut the lines away, use a craft knife or linoleum cutting tools. The craft knife works well, but you will need to make two cuts in order to make a "V" cut in the tile surface (see illustration). Very thin lines as well as much wider areas may be cut away. *Keep your fingers out of the way of the blade!*

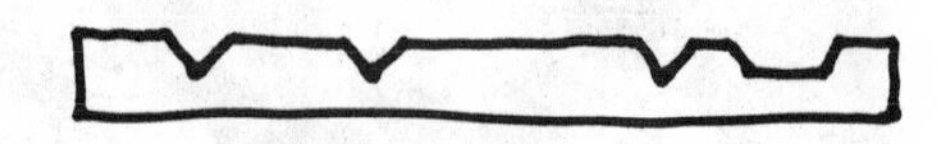

Printing. Cover the work surface with newspapers. Squeeze about an inch of printing ink onto a styrofoam meat tray or another linoleum tile. Roll it out in all directions using a rubber brayer (roll until the brayer is evenly covered with ink). The object of the rolling is to get the rubber brayer evenly covered, not the tray or tile.

Now, roll the inked brayer over the cut linoleum tile printer. Keep rolling until the entire printing surface is evenly covered with ink.

Place the tile on clean newspapers and then place a piece of paper over the inked surface. Rub the entire back of the paper to be sure all of the design is clearly transferred.

Peel the paper off the tile and re-ink for another print. Suggest to your pupils that they sign their names on prints and give them to friends as gifts.

Tubes & Bulbs

Light with a Beat

These papier-mâché instruments shake, rattle, and roll—beating out light-hearted music from south of the border. Children of all ages will enjoy making and playing these light bulb maracas.

MATERIALS

burnt-out light bulb
facial tissue
newspapers
white glue (diluted 50% with water)
brushes
tempera paint or acrylic paint
shellac or acrylic medium

PROCEDURE

Tear (don't cut) newspapers into strips about ½″ wide. Cover the light bulb with a sheet of facial tissue that is smoothed on with a brush and water. This will keep the newspaper strips from sticking to the bulb.

Cover the entire bulb with strips of newspaper, brushing the glue mixture on to secure all the edges. Continue the process until the bulb is covered with at least four layers of strips.

Then, cover the whole thing with a layer of facial tissue and glue. This will give a good surface for painting.

Allow to dry on waxed paper to avoid sticking. After it is thoroughly dry, hit it on a hard surface to break the bulb and thus create the rattle.

Paint with tempera paint or acrylic paint. If tempera is used, a coat of shellac or acrylic medium is advisable to make it more durable.

Light Bulb Puppets

The puppet characters designed with this form of papier-mâché are very versatile and can be made and used by children of all ages. A burnt-out light bulb is used as a mold on which the head of the puppet is fashioned.

MATERIALS

burnt-out light bulb
newspapers
white glue (diluted 50% with water)
paper towels
brushes
fabric
needle and thread
tempera or acrylic paint

PROCEDURE

Tear (don't cut) newspapers into strips about 1″ wide. Using a brush, cover each strip with diluted white glue and then apply it to the light bulb. Cover only the glass part of the bulb, leaving the metal part exposed. Smooth the strips out with the brush and glue.

Continue wrapping the bulb until it is covered with three layers of strips.

Then make the raised features, such as eyebrows, nose, and mouth. This is done with newspaper or facial tissue saturated with the glue mixture. It is easy to form if it is mixed until it is like clay.

Now cover the whole thing with more strips ½″ wide over the details to avoid wrinkles.

A final coating of paper towel strips will give a good surface for painting. Allow the head to dry thoroughly.

Carefully hit it on a hard surface to break the glass bulb inside. Empty the broken glass out of the open bottom and dispose of it in a trash bag or box.

Paint the head with tempera or acrylic paint. Add other materials, such as yarn for hair or a paper hat.

Costume. Fold a piece of fabric (about 10″ x 14″) and stitch along the edge to form a tube (see Illustration A).

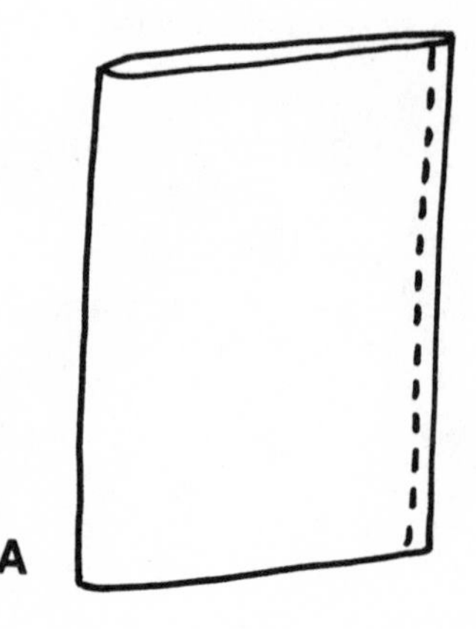

Run a gathering stitch around the top of the fabric to fit the neck of the puppet (see Illustration B).

Apply undiluted glue around the neck of the puppet and slip it into the costume. Pull the drawstring and tie securely.

Add a collar or bow.

Cut hands out of paper or fabric and sew them onto the costume (optional).

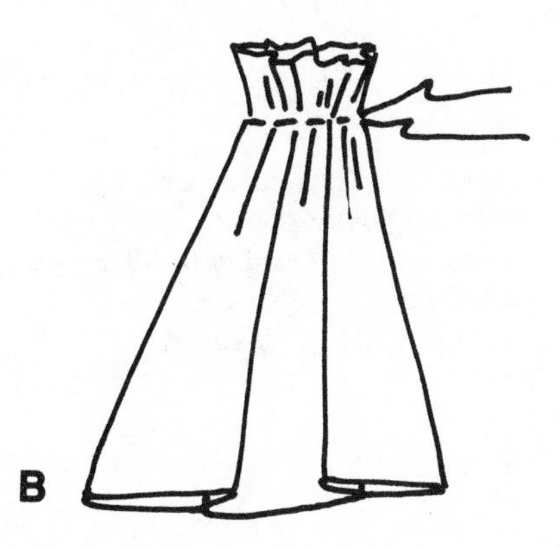

Tree Ornaments

Your students will enjoy making these decorative bulbs for the Christmas tree. These beautifully different ornaments will enhance any tree and make all eyes light up.

MATERIALS

discarded light bulbs and flashbulbs
colored tissue paper
white glue (diluted with 50% water)
brushes
beads, glitter, or sequins

PROCEDURE

Tear (don't cut) brilliantly colored tissue paper into small pieces. Apply the tissue to the bulb with a brush and the glue mixture. Smooth each piece out until the whole bulb is covered. Glue a loop of yarn or string on the screw end of the bulb (see illustration). Glue strips of tissue around the yarn ends and hang to dry thoroughly.

Additional enrichment may be added by using pure white glue to glue on sequins, beads, glitter, or metallic thread.

Flashbulb Robot

An amusing robot can be made with a flashbulb head on a box body. Small gift boxes or individual-serving cereal boxes are just about the right size for use with flashbulbs or flashcubes. Your class could make a whole series of creatures with square bodies and heads.

MATERIALS

flashbulbs or flashcubes
small boxes
tempera paint
felt pens
cardboard scraps
white glue (diluted with 50% water)
facial tissue
scissors
brushes

PROCEDURE

Cut a hole in the top of the box just a little smaller than the base of the bulb. Force the base of the bulb to snap into the hole. Cut arms and legs from scrap cardboard and glue them to the box.

Using white glue diluted 50% with water, cover the bulb with torn strips of tissue. Bring the strips on down to the box from the bulb. This will help to hold the head in place securely.

Allow to dry thoroughly.

Proceed to paint the box-creature with tempera paint—to create a person, robot, or whatever. Add details with felt pens after the paint has dried.

Other materials might also be added for enrichment, such as fur, hair, foil, or cloth.

Towers of Tubes

Cardboard toilet tissue tubes are great building blocks for the would-be sculptor or architect. New, free designs may grow from the hands of an inventive youngster as he or she stacks, cuts, and glues the tube shapes together.

Perhaps they will be nonobjective sculptures or even futuristic designs for buildings. Whatever they are called, the experience will be a rewarding investigation of space. Children of any age who work with patience and care will succeed in creating interesting constructions.

MATERIALS

cardboard tubes (from toilet tissue, etc.)
white glue
scissors
tempera paint or spray paint

PROCEDURE

Start by stacking tubes to create a strong base. Apply glue to one tube and press it against another. Support the glued tubes until completely dry.

Cut tubes may be used along with full-sized ones. This will add visual variety to the construction. Cutting slits in the tubes so they will lock together will also add variety, and at the same time add strength to the piece as well.

When completed, the constructions can be painted. One color will unify the design and keep the emphasis on shape and light and dark patterns. However, more colors could be used if additional color interest is desired.

If spray paint is used, be sure to do it outdoors and protect the area being sprayed. The best way is to set the entire construction inside a large box and hold the sprayer within it.

Alternative: Use cardboard rolls from tape, such as masking tape and transparent tape. Proceed in the same way as with the tubes. Cutting, however, is more difficult, since the cardboard is very heavy. To eliminate this problem, just use the rolls in their full size without cutting them.

Napkin Rings

Cardboard tubes from toilet tissue, waxed paper, and paper towels make a good base for papier-mâché napkin rings. One tube will make enough rings for a whole family. Your class will find this project easy and fun. Papier-mâché is a popular craft material suitable for use by younger and older children alike.

MATERIALS

cardboard tubes
white glue (diluted 50% with water)
yarn scraps
facial tissue
waxed paper
newspapers
brushes
scissors
tempera paint or acrylic paint

PROCEDURE

Cut 1½″ to 2″ rings from the tubes using heavy scissors.

Tear (don't cut) newspaper strips about ½″ wide. The torn edge works much better than a sharp, cut edge.

Apply the 50% glue mixture to the strips and wrap them around the cardboard ring. Cover each ring with 2 or 3 layers. Apply a final coating of facial tissue torn into strips.

Line designs can be created by dipping the yarn scraps in the glue mixture. Squeeze out the excess glue by gently pulling the yarn between the fingers. Now lay it on the ring in the design you wish. Be sure there is contact all the way along the yarn line. To ensure contact, light pressure with a pencil or stick will be helpful. When the glue dries, the yarn will form a hard, raised design (the completed design should be allowed to dry on waxed paper so it won't stick).

Painting. Creating an antique effect is accomplished by applying two coats of different colors of paint. Use acrylic paint or add white glue to tempera paint for permanence. Add about four parts glue to one part paint.

Paint with a solid, dark color. Let it dry completely.

Dry brush with a contrasting light color. To achieve a dry brush, dip the bristle brush into paint and wipe most of it off on newspaper. Quickly brush across the raised design so the paint will stay only on the raised surfaces. This will create the antique look.

A final coating of clear acrylic or pure white glue will give added protection to the napkin rings.

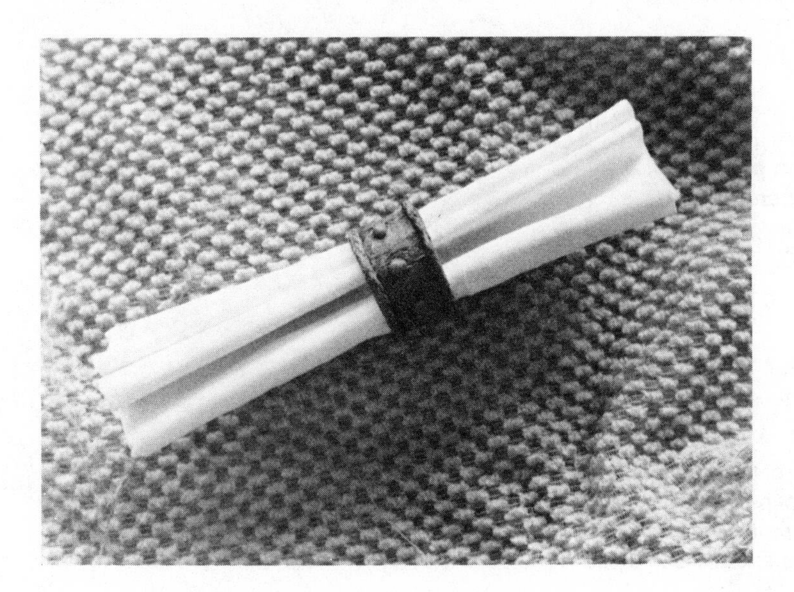

Noisy Birds

Your class can use these bird noisemakers made from cardboard tubes as rhythm instruments or for a rain dance. What fun it would be for a whole "flock" of these birds to get together in an improvisational dance!

MATERIALS

cardboard tube
cardboard roll from a pant hanger
white glue
tempera paint
brushes
felt markers
scissors
colored construction paper
yarn scraps or feathers
dried beans, peas, or gravel

PROCEDURE

Glue the cardboard roll from a pant hanger to the inside of a toilet tissue tube. Place it just to the top edge of the tube.

Trace around the edge of the tube to make a circle on a piece of construction paper. Draw several tabs, about ½″ long, all the way around the circle (see Illustration A). Cut it out and make another one by tracing around the one just cut.

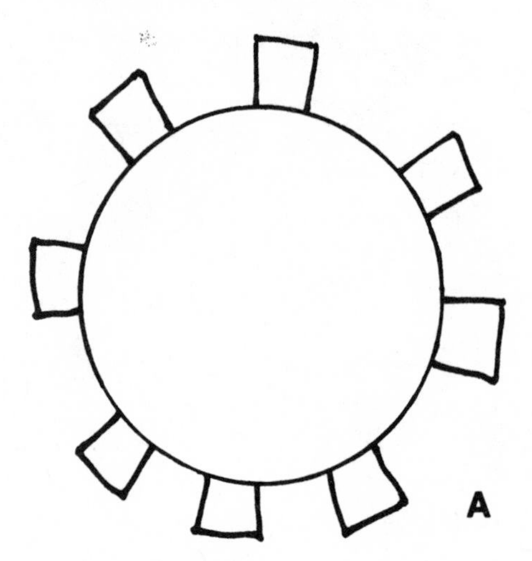

Glue one circle on the top of the tissue tube by applying a small amount of glue to each tab (see Illustration B).

Put dried beans, peas, or gravel (six to eight pieces) inside.

Glue the other circle on the bottom. The cardboard stick will be in the way at the bottom, so you'll have to fit the circle around it. It needs to fit tightly, so a little masking tape around the stick may be of some help.

Cut a beak out of construction paper and glue it to the front of the head (tube).

Decorate with eyes, feathers, etc., using tempera paint or felt markers.

Glue yarn or feathers to the top of the rattle for a plume.

Have everyone shake these bird-rattles as they dance, or use them to keep time to music.

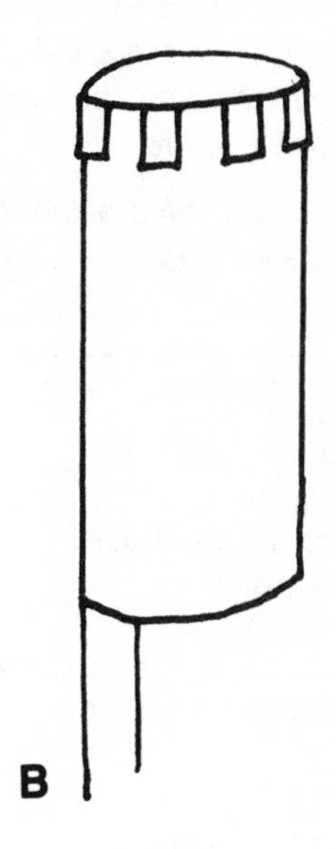

Inner Tube Prints

Old inner tubes from car or bike tires play an essential role in this printmaking project. Have your class ask parents and neighbors for discarded inner tubes. There comes a time when a tube cannot be repaired—and that's the time to consider it for making a printing plate.

This is a build-a-block method. The printing plate is cut and put together, rather than cut away as in a woodcut or linoleum cut.

MATERIALS

inner tube
scissors
heavy cardboard
printing ink
newspapers
rubber brayer (ink roller)
paper for printing

PROCEDURE

Cut shapes from the inner tube and mount them with rubber cement on a very heavy piece of cardboard. Any size pieces and shapes may be used, but don't overlap them. Keep the printing plate flat.

Printing. Cover the work surface with newspapers. Squeeze about an inch of printing ink onto a glass or plastic tray. Roll it out in all directions using a rubber brayer. Roll until the rubber is completely and evenly covered. The object is to cover the brayer evenly, not the glass or tray.

Now, roll all raised pieces of inner tube with the inked brayer. Roll back and forth slowly, until the printing plate is evenly covered.

Place the paper *on the printing plate*. Rub the back of the paper with the thumb to get a good sharp print. Carefully peel the paper off the printing plate. Re-ink the plate and continue to pull prints. Sign each one before hanging.

This technique works very well for printing greeting cards.

Rocks & Seeds & Shells & Weeds

A Colorful Paperweight

Set your students to searching for interesting, simply shaped rocks to be decorated with paint for unusual paperweights. They can give these paperweights as gifts or keep them for themselves. As gifts, they'll be appreciated by anyone.

MATERIALS

smooth rock
acrylic paint (or tempera with shellac coating)
brushes

PROCEDURE

After selecting the rock to be painted, wash and dry it thoroughly. Do some sketches of possible ideas for decorating the rock. Faces, animals, flowers, or butterflies will transform the rock into a handsome paperweight.

Acrylic paint is best because it is more permanent than tempera paint and easier to clean up than enamel. If tempera paint is used, it should be coated with shellac or clear acrylic after it is completely dry.

Another alternative is to mix a little white glue with the tempera (about a teaspoon per two tablespoons of paint) so it will adhere better.

Eggshell Mosaic

A mosaic is a composition made of small pieces of material placed next to one another. Eggshells that have been crushed and colored can be used to create an interesting mosaic effect.

Have class members begin collecting eggshells several days before the start of the project. After Easter is a good time for this, but it is certainly not limited to only that time of year.

MATERIALS

eggshells
watercolors or food coloring
white glue
heavy paper (e.g., shirt cardboards)
brushes

PROCEDURE

Crush the eggshells into very small pieces. If they stick together, peel off the membrane from the inside of the shells.

To color the shells, put them in a cup, pan, or jar and stir with a brush loaded with watercolor. Keep adding color until they are the color desired. Repeat the process for each different color you need.

Food coloring can be used instead of watercolors. In fact, if the shells are collected after Easter, many of them will already have food coloring on them.

Alternative method: If large quantities of shells are going to be colored at one time, dissolve watercolor in jars of water and allow the shells to soak for several minutes or longer. Food coloring may be used with this method also.

Draw a design on a piece of heavy paper, such as a shirt cardboard. A felt pen may be used if you want a heavy line to show. If not, use a light pencil line. Apply white glue to small areas at a time. A brush will allow you to control the glue better than using it directly out of the bottle.

Sprinkle on the crushed eggshells—one color at a time. Lay a piece of paper on top of the shells and press them into the glue. Let them set for a few minutes and shake off the excess. Add more glue and continue until the design is complete. Be sure to let the glue set before shaking off the excess shells.

The background may be left uncovered (as in the example shown) or it can also be covered with shells.

For added protection and adherence, a coating of acrylic or lacquer spray could be applied, but it is not necessary.

Rocky Bookends

Have your class collect sets of heavy rocks with interesting shapes. These rocks can be painted with representations of people, animals, or flowers and used as bookends. Simply coating the rocks with clear acrylic or shellac will give them a shine and intensify their colors. They make excellent gifts for Father's Day or Christmas.

MATERIALS

acrylic paint (or tempera paint with shellac coating)
brushes
felt
white glue
rocks

PROCEDURE

Have each pupil bring in a pair of heavy rocks that have an interesting shape and/or color design. Make sure they are flat enough so they will sit level for holding up books.

Have students decorate their rocks with acrylic paint or tempera paint. If tempera paint is used, be sure to coat the rocks with shellac or clear acrylic medium after the paint is dry. Or, if the natural colors of the rocks are to be retained, simply give them one or two coatings of shellac or clear acrylic.

To protect furniture surfaces, have the children glue pieces of felt to the bottom of the rocks.

Peter Cottontail's Easter Eggs

Your class can compete with Peter Cottontail by decorating their own eggs. They can use hard-boiled eggs or blown-out eggshells—the latter can be saved for years to come. Children of all ages will enjoy creating these delightful spring goodies.

MATERIALS

whole eggshells (or hard-boiled eggs)
permanent ink felt markers
pin or small nail

PROCEDURE

If the egg is going to be blown out, put a pinhole in each end. Make the one in the broad end a little larger. Hold the egg over a bowl and gently blow into the small hole until the eggshell is empty.

Carefully wash and dry the hollow shell.

Permanent ink felt markers come in beautiful colors and will make the designs drawn on the eggs last for years. A coating of clear acrylic spray may be used to further protect the egg.

Egghead s

Creatures of all kinds can be made with eggshell heads. Pupils can make characters from a story, people out of history, clowns in a circus, or ghouls at Halloween. This project is suitable for all ages as long as the children realize that eggshells are fragile and must be handled with care.

MATERIALS

whole eggshell (empty)
permanent ink felt markers
colored construction paper
scissors
white glue
pin or small nail

PROCEDURE

Put a pinhole in each end of the egg. Make the one in the broad end a little larger. Hold the egg over a bowl and gently blow into the small hole until the shell is empty. Gently wash and dry the shell.

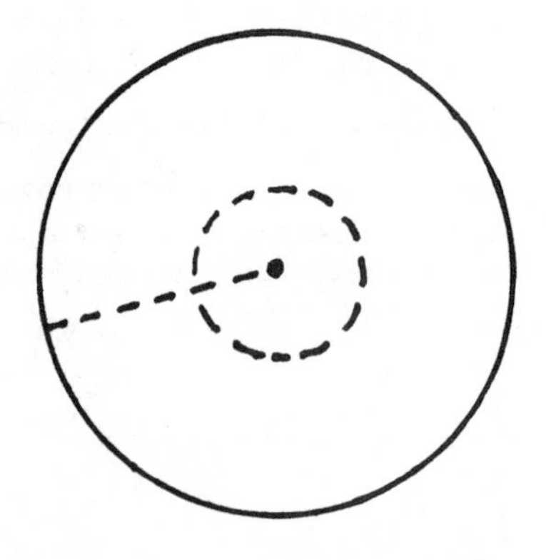

To make the neck for holding the egg head, cut a circle out of construction paper (about 6″ in diameter or more). Cut from the edge of the circle straight into the center. Cut another circle out of the center of the big circle (see Illustration). Overlap the paper to create a cone and glue with white glue. The hole at the top of the cone will be where the egg head will rest.

Have the children draw faces, hair, etc., onto the eggshells using permanent ink felt markers. Hair could also be added with yarn or fur, and eyeglasses of wire might even be used. Encourage the children to be imaginative!

Have them set each completed head in its paper cone, to which they might add a collar, necktie, brooch, and so on. Add a little white glue to the top edge of the opening in the cone to secure the egg.

Rock Relief Sculpture

Little rocks of different sizes, shapes, and colors need to be collected for this project. They can be gathered during a field trip to the woods, the park, or the beach. Keep them natural this time (do not paint them), relying only on their shapes and arrangement for interest.

MATERIALS

wood backing (scrap is okay) or heavy cardboard
tacky craft glue
twigs
pull ring from soft drink can

PROCEDURE

Have each child look at his or her collection of rocks. Tell the children to move the rocks around and to group them in different ways. Do they suggest the shape of a bird, an animal, a person, or a flower?

Suggest using some twigs along with the rocks for adding lines to the design.

When the arrangement is satisfactory, begin to glue the rocks onto the piece of wood. An irregular piece of scrap wood will do very well. If wood isn't available, very heavy cardboard may be used.

Staple or glue a pull ring from a soft drink can to the back for a hanger.

Don't Plant It—Glue It!

Watermelon seeds, pumpkin seeds, and many other kinds of seeds make visually interesting compositions. All of these can be glued to a surface to create an effective mosaic-like panel.

MATERIALS

seeds
white glue
heavy paper

PROCEDURE

Wash and dry the collected seeds.

Make a very light pencil sketch on the heavy paper to act as a guide in building the mosaic. Pictures of subjects such as birds, fish, flowers, or insects might be used for motivation. Apply glue to small sections at a time.

Lay the seeds on the wet glue as though painting with brush strokes.

Different kinds of seeds, or just one type, may be used on one picture. One of the examples shown is made of watermelon seeds, while the other has a variety of seeds and beans.

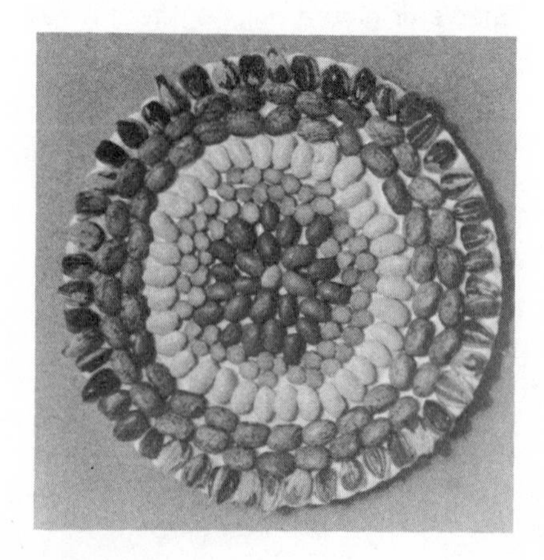

A Note on Nature

Beautiful note papers seem to grow as this project develops. It all begins on a nature walk—with a search for ferns, small leaves, and flowers. What a nice way for your students to say they care—by sending cards they have made themselves!

MATERIALS

natural leaves, ferns, flowers
construction paper or similar
white glue (diluted 50% with water)
bristle brush
facial tissue

PROCEDURE

Choose ferns, leaves, and flowers that will remain flat on the paper.

Cut the paper to fit a standard-size envelope or the class will have difficulty using their creations. Fold the paper in half like a card.

Glue the natural materials to the front of the card using undiluted white glue. Place a heavy book on top to keep them flat as the glue dries.

Tear the edges off the facial tissue to remove the hard, sharp edge. Lay it on top of the natural objects.

Gently tap over and around the leaves and petals with the tip of a bristle brush saturated with diluted white glue. The glue dries clear, so what seems to disappear as you work will reappear as the glue dries.

Allow to dry thoroughly.

Rub Out the Weeds

Frottage *is a French word for rubbing. This is the technique used in this project to transfer the designs of weeds and grasses onto paper. The search for weeds, ferns, and grasses is a good nature experience, and arranging them for the* frottage *is a good art experience.*

The visual awareness developed through finding, feeling, and smelling is a bonus to the esthetic experience.

MATERIALS

natural grasses, weeds, etc.
crayon
lightweight paper

PROCEDURE

Go on a nature walk to gather a variety of natural materials, which must be relatively flat.

Break a peeled crayon so it is no longer than 2″.

Arrange some of the grasses, weeds, or ferns on the table and lay a piece of lightweight paper on top of them. Hold the paper flat and secure it by spreading your whole hand out over it. This will prevent the natural pieces from moving during the rubbing.

Holding the crayon flat on the paper, slowly make circular motions over the paper. Gradually the shapes of the objects will appear. Don't try to press too hard. It's better to build up the shapes a little at a time. Get as much of the detail to show up as possible, such as the veins in the leaves.

The objects can then be rearranged and the process repeated.

It Grows on You

Your class will enjoy making jewelry with seeds. Any time a watermelon is eaten, there are hundreds of seeds to dispose of, and what better way than to string them into a necklace or bracelet! Several strands would make a very striking piece of summer jewelry.

Other kinds of seeds can also be used. Combine them! Experiment!

MATERIALS

seeds (watermelon, pumpkin, apple, etc.)
small needle
strong thread
small safety pin

PROCEDURE

Thread a small needle with strong thread, such as nylon. Double it and tie a knot.

Clean off the seeds.

Push the needle through the softest part of the seeds. If the seeds are very brittle, it helps to soak them in water for a while before stringing.

Tie one end of the thread to a safety pin and make a loop in the other end for fastening.

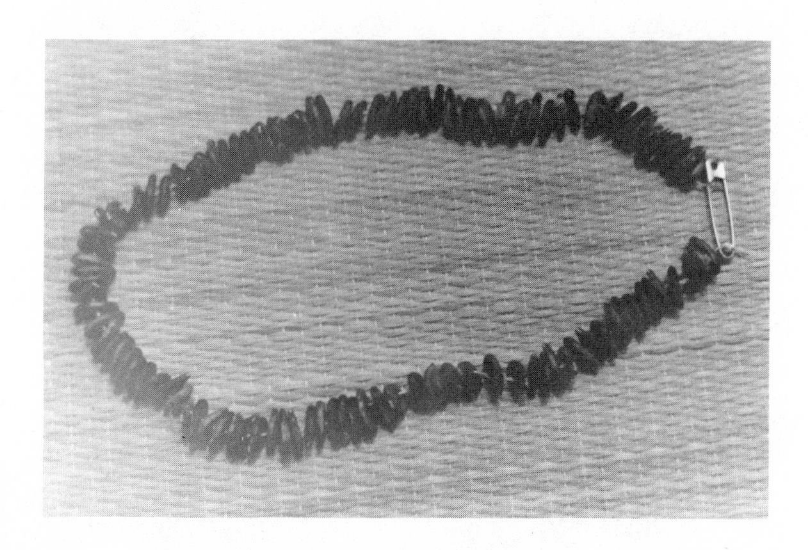

Candles & Corks

Candle Drop Beads

Partially used candles that drip can be used to make little wax beads for stringing into necklaces. The colors may be varied or just one can be used. The resulting necklace resembles paper shell "hishi." Be sure to use extreme caution when using an open flame. It is best to reserve this project for older children.

MATERIALS

used candles (not the dripless kind)
waxed paper
thread
very small needle
newspapers

PROCEDURE

Spread a piece of waxed paper about 12″ long on newspaper.

Light the candle, which should not be the dripless type for best results. Tip it on its side and hold it about 1″ or 2″ above the waxed paper. Make a series of rows of wax dots as close together as you can, being very careful not to touch the waxed paper with the candle. Continue until the paper is covered with dots of wax.

Stringing. Thread a *very small* needle and double the thread (a little longer than the finished length will be). Tie a knot in the ends.

Push the needle into the center of each dot with the paper raised slightly. Remove them from the waxed paper. Two or three dots can be pierced onto the needle at one time. Then put your fingernail against the side of the needle to slide the dots onto the string. Do this carefully to avoid breaking the wax. Some will crack, so just discard them.

When all of the dots are used, make more. Then continue to string.

For variety, wooden or paper beads could be interspersed among the wax ones.

Avoid getting the completed necklace in direct heat —for obvious reasons!

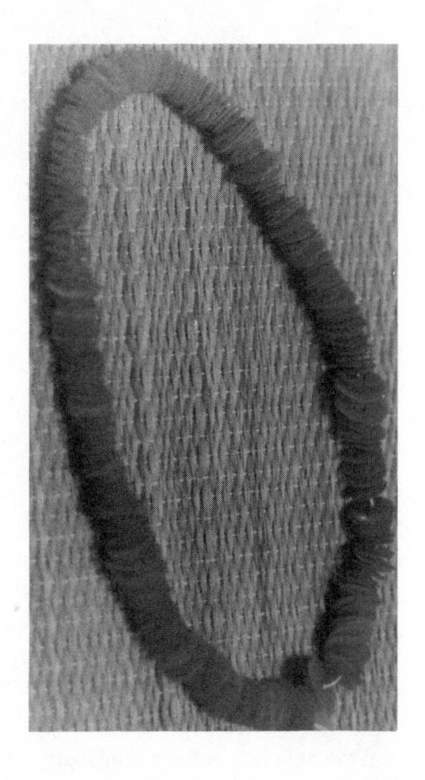

A Little Drippy Drawing

Encaustic painting dates back to the Egyptians. Painting with wax and pigment has been done in many parts of the world and is still done today. This activity is actually an imitation, or pseudo-encaustic, using broken crayons and partially burned candles.

Close supervision is advised while using the lighted candles due to the obvious dangers of fire. Keep a container of water nearby as a precaution.

MATERIALS

used candles
newspapers
heavy paper or cardboard
broken crayons
container of water
old spoon

PROCEDURE

Lay heavy paper or cardboard on newspapers. (Lightweight paper is too pliable and the wax will chip off.)

Set the candle in a holder and light it. Be very cautious when working with the flame.

Hold the crayon over the flame just close enough to cause it to melt. As it melts, hold it over the cardboard or touch it to the cardboard surface.

When the pieces of crayon are too small to hold over the flame, place them in an old spoon and hold it over the flame. Use a ¼″ strip of cardboard to dip the melted crayon out of the spoon and onto the design.

Work for light and dark contrasts in your design. It can be representational or purely an optical experience in color.

Alternative. Decorate the lid of a box with this technique to create a very interesting collector's box. It might be used to store crayons.

Cork Prints

A cork of any size has two flat ends that can be cut into a design for printing. Your class can use carved corks to print a colorful repeated design on a book jacket, wrapping paper, or even fabric. Children of all ages will find pleasure in this activity, and at the same time will learn about the beauty of repetition and all-over pattern.

MATERIALS

corks
hacksaw blade, jackknife, or craft knife
paper or fabric for printing on
tempera paint or fabric paint
pan for paint (e.g., TV dinner tray)
brushes
newspapers

PROCEDURE

An old hacksaw blade can be broken into several pieces and used as a tool for cutting the cork. This makes a very safe tool for young children to use. If this is not available, any kind of knife can be used—*with caution*.

Cut a design into the surface of both ends of the cork. This will give you two designs to print from on one cork.

Restrict the designs to straight lines since curves are very difficult to cut due to the nature of the material.

Cut the lines at angles so you'll actually be cutting a "V" shaped section out (see Illustration A).

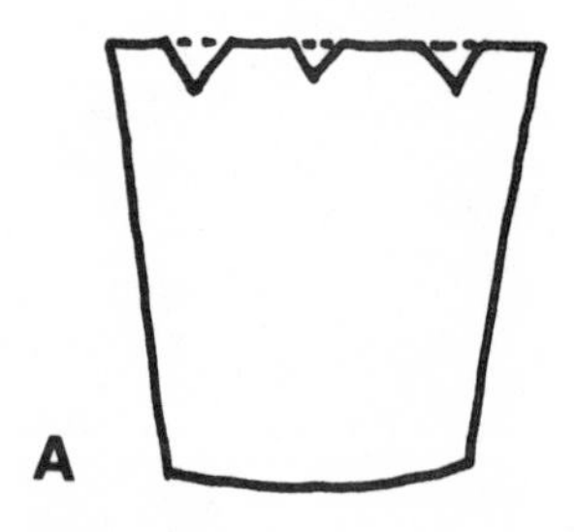

Printing. If the printing is being done on paper, use tempera paint. For fabrics, use paint designed for permanence on textiles and follow the directions on the label.

Cover the table with several layers of newspapers to provide a soft cushion.

Pour paint into a TV dinner tray or similar type of container.

Brush paint onto the cork so the raised surface is evenly covered with color.

Press the cork onto the printing surface. The two designs, one on each end of the cork, may be alternated in a regular or irregular pattern. Two of many possibilities are shown in Illustration B.

For a clear print, apply paint each time a new print is made.

X O X O X
O X O X O
X O X O X
O X O X O

B

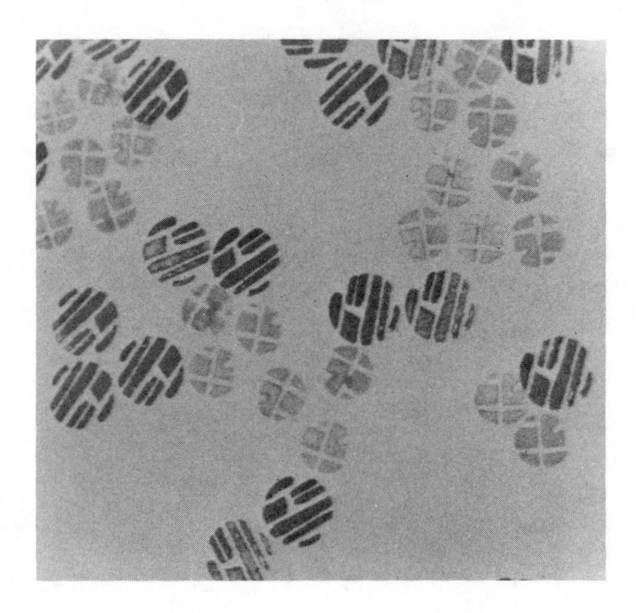

Magic Drawings

Sometimes called ghost drawings, these line drawings are difficult to see until the final step—when they seem to appear like magic. The excitement and mystery of Halloween will be heightened with this activity, but it is certainly not limited to only that time of year.

Watch little eyes become large with amazement as the lines appear under the paint brush.

MATERIALS

used candles
white drawing paper
watercolor or thin tempera
brushes
newspapers

PROCEDURE

Use a candle to draw a picture on a piece of white drawing paper. Press hard so a heavy coating of wax will be transferred to the paper.

Place the completed drawing on a newspaper.

Brush very thin watercolor or tempera paint over the whole surface. Be sure the paint is watery or it will cover the lines. The wax will resist the watery paint, allowing the candle wax lines to show through. One or several colors may be used for these intriguing drawings.

Corky Creatures

Your class will have fun making cork animals. Mini-menageries make table decorations that are amusing and full of character. In a cage or running wild, these are animals everyone will enjoy making.

MATERIALS

corks
hacksaw blade or knife
pipe cleaners
nail
felt pens

PROCEDURE

Your pupils will need a cork for the body of each creature. Have them make a nail hole in each appropriate place for legs, tail, and neck.

Cut pieces of pipe cleaners for these added appendages.

Carefully cut another cork, using a hacksaw blade (broken or discarded) or a knife, to form a head. Put a nail hole in it and attach it to the body with a piece of pipe cleaner which will represent the neck.

Now have the children create designs on the body and a face on the head of each creature—using felt pens and lots of imagination! These mini-creatures don't have to look real. In fact, they're more fun if they don't.

Old gift boxes might be cut and painted to give the animals a cage or a home.

How about decorating a food tray for a shut-in with a sprightly little animal that looks up and says "get well!"

Head Pendant

This head pendant, made from a cork, can be worn on a chain or on a string. Little girls may want a dainty little face while a boy may want a warrior. Even birds or animals could be used. Ask your class for other ideas.

MATERIALS
cork
permanent ink felt pens with fine points
yarn
sequins, map tacks, small buttons, beads or feathers
chain or string for hanging
tacky craft glue
wire or paper clip
wire cutters

PROCEDURE

Choose a cork of appropriate size. Draw a face using permanent ink felt pens with fine points. The eyes might be sequins (as in the example shown), colored map tacks, small buttons, or beads. Glue these on with tacky craft glue.

Glue on hair or feathers to suit the type of character being designed.

Push a wire loop into the top of the head. This can be made from heavy wire or by cutting the end off a paper clip. Have the loop going from front to back so the pendant will hang straight (see illustration).

Index

Index

Index

Index